MORAL DILEMMAS IN REAL LIFE

Law and Philosophy Library

VOLUME 74

Managing Editors

FRANCISCO J. LAPORTA, *Department of Law,*
Autonomous University of Madrid, Spain

ALEKSANDER PECZENIK†, *Department of Law, University of Lund, Sweden*

FREDERICK SCHAUER, *John F. Kennedy School of Government,*
Harvard University, Cambridge, Mass., U.S.A.

Former Managing Editors
AULIS AARNIO, MICHAEL D. BAYLES†, CONRAD D. JOHNSON†,
ALAN MABE

Editorial Advisory Board

AULIS AARNIO, *Research Institute for Social Sciences,*
University of Tampere, Finland
ZENON BAŃKOWSKI, *Centre for Law and Society, University of Edinburgh*
PAOLO COMANDUCCI, *University of Genua, Italy*
ERNESTO GARZÓN VALDÉS, *Institut für Politikwissenschaft,*
Johannes Gutenberg Universität Mainz
JOHN KLEINIG, *Department of Law, Police Science and Criminal*
Justice Administration, John Jay College of Criminal Justice,
City University of New York
NEIL MacCORMICK, *European Parliament, Brussels, Belgium*
WOJCIECH SADURSKI, *European University Institute,*
Department of Law, Florence, Italy
ROBERT S. SUMMERS, *School of Law, Cornell University*
CARL WELLMAN, *Department of Philosophy, Washington University*

MORAL DILEMMAS IN REAL LIFE

Current Issues in Applied Ethics

by

OVADIA EZRA

Tel Aviv University, Israel

 Springer

A C.I.P. Catalogue record for this book is available from the Library of Congress.

ISBN-10 1-4020-4103-9 (HB)
ISBN-13 978-1-4020-4103-7 (HB)
ISBN-10 1-4020-4105-5 (e-book)
ISBN-13 978-1-4020-4105-1 (e-book)

Published by Springer,
P.O. Box 17, 3300 AA Dordrecht, The Netherlands.

www.springer.com

This book was published with the support of The Minerva Center for Human Rights
in the Buchmann Faculty of Law at Tel Aviv University

Printed on acid-free paper

All Rights Reserved
© 2006 Springer
No part of this work may be reproduced, stored in a retrieval system, or transmitted
in any form or by any means, electronic, mechanical, photocopying, microfilming, recording
or otherwise, without written permission from the Publisher, with the exception
of any material supplied specifically for the purpose of being entered
and executed on a computer system, for exclusive use by the purchaser of the work.

Printed in the Netherlands.

CONTENTS

Acknowledgements	vii
Preface	ix
A. INDIVIDUAL RIGHTS AND PUBLIC DUTIES	**1**
1. Privacy and the Public Sphere	5
2. The Obligation of the State toward Individuals	15
3. Public Security vs. the Right to "Be Let Alone"	25
4. Freedom of Expression in Academia and the Media	37
B. MEDICAL ETHICS	**51**
5. Mercy Death or Killing	55
6. Donating or Selling Organs	69
7. Genetic Engineering and Reproduction	81
C. PARENTHOOD AND THE FAMILY	**97**
8. Rights of Relatives and Generations	101
9. Procreation after Death	115
10. Babies as Commodities	127

D. PUNISHMENT — **141**

11. Punishment of Sex Offenders — 145

12. Punishment and Domestic Violence — 157

13. Capital Punishment and the Mentally Retarded — 169

Index — 183

ACKNOWLEDGEMENTS

I would like to express my deep gratitude to some of those who have helped me in writing this book.

Dr. Zvi Tauber, for his assistance in forming the structure of this book and for his helpful comments on each chapter of it.

Prof. Anat Biletzki, for her help in forming the proposal of the book, and for her encouragement and spiritual support during the writing.

Prof. Jacob Joshua Ross for his comments on chapter 1 and 4.

Prof. Daniel M. Cook for his Comments on chapters 2 and 3.

Mrs. Sandy Bloom, for the wonderful job she did in editing, despite the impossible time table.

To the Minerva Center for Human Rights in the Buchmann Faculty of Law at Tel Aviv University, for their financial support.

PREFACE

Life presents both individuals and communities with situations that demand moral and ethical deliberations. From the more general issues of universal globalization to the very specific problems of everyday existence encountered by active agents, contemporary life is replete with moral and ethical conundrums. Any thinking person is required, so it seems, to be concerned, involved, or—at the very least—conversant with the ins and outs of ethical argument. This book purports to supply ways of thinking of, perhaps even dealing with, such argument.

Applied ethics is that intellectual locale where theory meets praxis; this book is designed to make that meeting point explicit, by presenting parts of that series of issues in well-grounded philosophical formulations. As such, it hopes to serve both the academic milieu and the intellectual lay public.

The format of the book fulfills the above agenda. It begins with the general relation between the individual and society—that topos which instills ethical tension, and even conflicts, between the private and the public in our discourse. The book then segues from general to specific as it gradually narrows the ethical playing field and relates to medical ethics, the family, and the practice of punishment. In all cases of dealing with the ethical "issues" in our list, the book addresses both consensual and conventional social institutions and distortions thereof. As such it opens up the area of discussion not only to theorists and intellectuals, but also to therapists, practitioners, and other professionals who look to ethics as a "form of life".

Before presenting the topics and structure of this book, I want to make a few comments on ethical discussions in general, and on these within the domain of applied ethics in particular. Ethical discussions are usually divided into three different levels. The highest one is called *Metaethics*. This level is that branch of ethics which seeks to understand the nature of ethical evaluations and moral statements. It inquires into the issues of the language of ethics, the use of this language, and sometimes even examines the foundations of morality, and asks whether it has objective validity or transcendental sources. Some theorists consider even the question of how to manage ethical discussions as belonging to this domain.

The next level of discussion is usually called *Normative Ethics* to which the well-known great and comprehensive ethical methods or ethical systems belong. This domain deals with the moral principles and rules, according to which a person should live and interact with other human beings. This domain has three main essential systems: deontological (which deals with moral duties and imperatives), teleological (or consequential, which deals mainly with future utility and happiness), and virtuous (which deals with moral virtues). Such ethical methods are supposed to be universal, but there are some which are not (such as relativist ethics and ethical egoism). In any event, normative ethics as a whole is supposed to provide an answer to the general question of how human beings should live.

The third level is *Applied Ethics*, the domain expressed in this book and which deals with the ethical aspects of specific issues. Its goal is not merely a theoretical discussion about the ethical aspects of those issues, but mainly attitudes or at least to cause people to re-assess them. Of course, any part of such a discussion leans on moral principles that were formulated by the previous domain; that is, it uses normative ethical theory in order to justify or validate its arguments. However, considerations in applied ethics can be taken simultaneously from more than a single normative ethical theory, if these different considerations do not conflict or contradict each other. What should be stressed is that each discussion in applied ethics is carried out within a comprehensive ethical, social, and political worldview, within which its presumption and arguments are valid or acceptable. Thus, its conclusion can be accepted only by those who accept the foundational principles upon which these arguments are grounded, while others, who reject that worldview, will reject the conclusions it dictates with regard to specific issues.

Accordingly, this book, whose form and structure are typical of the genre of applied ethics, starts from a deontological social democratic worldview, which tries to balance between the individual's freedom and social needs. It tries to maximize people's freedom but gives weight to considerations of social solidarity and equality. The book discusses issues that have occurred in real life and tries to present the moral principles and considerations that should be applied when dealing with such issues. Thus the specific examples, which were originally taken from the media, are only the triggers for discussion of the principal problems that are raised by these examples.

Most of the cases discussed in this book occurred in the United States, the United Kingdom, and Israel, mainly because the media in these countries is most accessible to me due to reasons of language and political culture. However, cases similar to these occur in any society, hence the discussions are relevant to most Western societies with slight modifications. After all, the issues of privacy, medical ethics, sex crimes, or domestic violence relate to people's lives wherever they live. The point is that since these countries share similar political and social attitudes, the discussion can be most relevant to their common political and social culture.

What is important to stress about the specific cases is their role in showcasing the principal problems of each chapter. In other words, the specific cases themselves are not the essence, but are only the springboard for discussion of relevant principles and considerations. For example: when discussing punishment, the principle of *lex talionis* (the law of retaliation which is a principle of retribution) is mostly relevant, and when discussing the specific punishment of mentally retarded offenders, the principle of *mens rea* or the criminal intent required to commit the crime) is of great relevance. Although the weight assigned to such principles can be variable, any discussion about such issues should at least mention and relate to them.

The classification of the different cases in the four sections of the book is not definitive. The issue of domestic violence, for example, appears in Section D that deals with punishment although it could also be discussed in Section C that deals with the family. The issue of procreating after death is discussed in Section C, even though it is also related to Section B that deals with medical ethics. It was more important to focus on the four general categories listed below (rights and duties, medical ethics, family, and punishment) than to rigidly attempt to categorize the cases used to illustrate the overall issues.

Similarly, many principles and ideas are discussed in depth in certain chapters and then are mentioned again in later chapters, due to their relevance. Thus, for example, many of the arguments that are presented in Chapter 1 about the significance of privacy, re-appear and are elaborated throughout Section A as a whole. The same is true for the arguments against the commodification and exploitation of human beings and their organs, which are raised in Chapter 6 (in the context of selling or donating organs), and re-appear in Chapter 10 (in a different section altogether) in the context of selling babies.

However, as said before, this book follows the usual format and structure of these in the genre of applied ethics. It contains four sections:

A. Individual Rights and Public Duties;
B. Medical Ethics;
C. Parenthood and the Family;
D. Punishment.

Each section includes three or four chapters and is dedicated to a particular problem or issue that is of public concern. I devote each specific chapter within the section to explicating the ethical uncertainties and complications of a current "headline" on the public agenda. I discuss the pros and cons of competing positions and solutions to the dilemma encountered therein, and finally offer a way of resolving the problem—even though, sometimes, imperfectly.

SECTION A. INDIVIDUAL RIGHTS AND PUBLIC DUTIES

This section deals with the inherent tension between the individual and the community. A person is both an individual as well as a citizen or a member of community, and this immanent dichotomy always involves conflicts.

These conflicts may occur between individuals who have their own preferences, desires, needs and interests, and the social framework within which the individual exists, which also has its own interests. The core of this conflict is the fact that many interests of the individual are acknowledged by society as valid claims against the social framework, and thus receive the status of *rights*. On the other hand, the individual, as a member of society, has obligations to fulfill toward the social framework, and thus is considered as having *duties*. Sometimes, the rights (of society) and the duties (of the individual) conflict with each other, and we have to decide which interest is more urgent and, therefore, should prevail.

Whenever there is a conflict between the individual's interests and those of the social framework, my own tendency is to support the individual's interests and provide them with more weight than the interests of the collective framework; this arises from a belief that the state or society is innately robust and can protect its interests more efficiently than the individual. However, this preference is not absolute and when the interests of the individual are only slightly harmed while the social framework is significantly benefited, then I accept the price of a minor violation of the individual's rights—but not without reservations.

This section deals with a few different domains within which such conflicts sometimes occur.

Chapter 1: Privacy and the Public Sphere

The first chapter deals with the problem of privacy and the public sphere from different points of view. It questions the immunity from, or privilege against, disclosure of private information accorded to the individual—public figures, ordinary people and even criminals—whereby it conflicts with the public interest. In this regard we deal with the question of medical confidentiality of public figures or elected representatives; exposure of the criminal records of minors who have grown up and want to be integrated into society; disclosure of names of public figures being investigated in criminal procedures, and so on. What I stress in this chapter is that there is a big difference between an ordinary person's claim for respecting his or her privacy, and the same claim when it is demanded by a public

Preface xiii

figure. I show how the issue of privacy becomes very complicated when it clashes with the public interest.

Chapter 2: The Obligation of the State Toward Individuals

This chapter deals with a more general topic: the obligation of the state toward individuals. Here we deal with the role of the state in protecting the welfare and personal security of deprived sections of society (including both citizens and residents). By "deprived" we mean children (or fetuses), mentally retarded individuals, or persons on the margins of society (minorities, fringe groups, etc.). We ask whether health and educational authorities are permitted to coerce individuals into normative forms of behavior, and whether they may desist from supplying services or basic necessities to such individuals. In this context we raise the question of the parental rights of mentally retarded people and the state's duties to them and their children. In this chapter I try to express an instrumental concept of the state and argue that the state is obligated to provide necessary means to its residents, even in cases where the original duty to provide such means was imposed on others (their parents, for example).

Chapter 3: Public Security vs. the Right to "Be Let Alone"

This chapter deals again with the problem of privacy, but from a different perspective: that of databases. This has become a burning issue since September 11, i.e. since "terrorism" made its way into public discourse as the most intimidating phenomenon to date. Thus for reasons of (purported) security, the state inquires into such issues as the question of the legitimacy of DNA data resources that may help to locate or identify criminals, and the question of trace and documentation of telephone, e-mail, and other means of communication. Our own statement, in light of the current political ambience, is that there is room for worry that authorities might abuse the populace's concern about security and violate basic rights in so doing. Here, as in the first chapter, I tend to accept the establishment of a very specific database, but with stringent restrictions on its use. When there is a demand for a comprehensive database, I outright reject it.

Chapter 4: Freedom of Expression in Academia and the Media

This chapter turns to different aspects of freedom of expression, which have characterized the ongoing debate in both academia and media. We ask about the academic freedom to express controversial views; for example, views that are couched in anti-feminist or anti-minority (call them racist) utterances in the classroom, on campus, in newspapers, on TV talk-shows, etc. A related issue, also addressed here, is the right (or duty) of the media to broadcast offensive or extremely gory events and images on television. We inquire whether it is either prohibited or compulsory to bring such images to public awareness. Here, too, even though I try to maximize freedom of speech, and allow "hate speech" at least in academia, I do not think that everything is worthy of being visually revealed to the public. I basically object to external censorship in the media, but in certain cases I expect the media to conduct self-censorship—not in verbal reporting but in televising brutal or gory scenarios.

SECTION B. MEDICAL ETHICS

Of the myriad of ethical debates going on in the present-day public discourse, it would not be an exaggeration to say that medical ethics raises the most intensive and vehement arguments. The rapid progress in technology and bio-technology has far outstripped parallel progress, if there be any, in either moral or legal studies. It seems that by the time that ethical discussion, buttressed by legal considerations, grasps and is able to deal with a medical issue—one that is usually on the frontier of scientific discovery—technological progress has already flung us forward into more complicated or acute issues. This section deals with some of these issues.

Chapter 5: Mercy Death or Killing

This chapter deals with the somber question of mercy death or mercy killing. Technological advances for extending human life and maintaining acutely-ill persons on life support may be perceived as both a blessing and a curse. Alongside new hope that is sometimes acquired by an extension of one's life-span, the preservation of a life gone awry often involves suffering for the patients and additional burdens for their caretakers. This chapter inquires, generally, into the legitimacy of both coercive treatment and abstention of such, for those who either refuse or cannot express their consent to medical care. A specific issue arising here, under the same theoretical umbrella, is the legitimacy of separating Siamese twins, when it is clear that at least one of them will die as a result.

Chapter 6: Donating or Selling Organs

This chapter deals with the more optimistic but no less sensitive and complicated issue of organ donation or organ sale. The possibility of saving life or enhancing its quality by organ transplantation, and the unwillingness of (sufficient) people to donate the required organs, raises a question about the morality of selling human parts. Two difficulties are immediately encountered: First, it is clear that poverty-stricken people are the ones most liable to risk their own health in order to procure funds. The problematics of commerce in human organs is revealed in the current chapter, most notably exemplified by kidney "donations." Secondly, and perhaps more marked by philosophical principles, the value-laden question of the reification and marketability of the human body is analyzed through the paradigm of ovum-contributions. The first is a problem of justice; the second— one of values.

Chapter 7: Genetic Engineering and Reproduction

This chapter deals with genetic engineering and reproduction. The field of genetic engineering encompasses the (insurmountable?) gap between technological progress and the inability of the ethical dictionary to respond to issues that arise in its wake. The most far-reaching debates in this area have to do with human reproductive cloning and stem-cell research. The conflicts between scientific demands and political or philosophical misgivings and qualms, sometimes make the issue intractable. We try to offer guidelines for therapeutic stem-cell research while, at the same time, banning human reproductive cloning, thereby attempting to address both sides of the dilemma consistently. A related issue, also discussed in this chapter, is the question of creating new siblings for the purpose of using their organs (usually bone-marrow) to medically assist their brothers and sisters. It is clear that future scientific/technological developments may, nevertheless, obligate us to rethink our own judgments—in this, as in all other questions in this section.

Much of this chapter deals with the danger of leaving such acute issues exclusively in the hands of the scientific community. This danger was clearly raised by Jurgen Habermas, whose book *The Future of Human Nature* discusses this in detail when he says: "The new technologies make a public discourse on the right understanding of cultural form of life in general an urgent matter. And philosophers no longer have good reasons for leaving such a dispute to biologists and engineers intoxicated by science fiction."[1] Genetic engineering challenges some of our most fundamental beliefs about morality. It enables us to control the physical basis which we are by nature, and as Habermas describes this problem, "What for Kant still belongs to the 'kingdom of necessity', in the perspective of evolutionary theory, changed to become a 'kingdom of contingency'." Genetic engineering is now shifting the line between the natural basis we cannot avoid and the "kingdom of ends." This extension of control of our "inner" nature is distinguished from similar expansions of our scope of options by the fact that it "changes the overall structure of our moral experience."[2]

---

[1] Habermas Jurgen. *The Future of Human Nature*. Polity Press. 2003. p. 15.
[2] Habermas Jurgen. *The Future of Human Nature*. Polity Press. 2003. p. 28.

One of the most fundamental changes of this kind, as Habermas understands it, may be the uprooting of the categorical distinction between the objective and the subjective, and this dedifferentiation "of deep rooted categorical distinctions which we have as yet, in the description we give of ourselves, assumed to be invariant. This differentiation might change our ethical self-understanding as a species in a way that could also affect our moral consciousness—the conditions, that is, of nature-like growth which alone allow us to conceive of ourselves as the authors of our own lives and as equal members of the moral community."[3]

However, genetic engineering might also have very promising consequences, particularly in the domain of stem-cell research. Stem-cells "can be used to repair organic damage, to recreate parts of the human body that are diseased or malfunctioning. Thus they present us with wonderful new therapeutic possibilities, several of which have already been impressively demonstrated. Bone marrow transplants to regenerate a healthy blood system in patients with leukemia, for instance."[4] This chapter deals with the dilemmas that are raised from these possibilities but its main theme is that we should restrict the research to therapeutic cloning and confine the immense potential of stem-cell research only to negative eugenics—all the while maintaining tight control over the researchers and scientists involved. Although this supervision should be, primarily, the mandate of the scientific community, the international community, and society as a whole should share the burden of monitoring the scientists. All of us should ensure that scientists do not cross the border between negative eugenics, which prevents diseases, and positive eugenics, which might genetically enhance the species which we know as homo sapiens, but with genetic enhancement might be changed into something else.

SECTION C. PARENTHOOD AND THE FAMILY

Modern—or is it post-modern?—life styles have created new forms of relationships between, within, among and outside the traditional nuclear family (consisting exclusively of mother, father, and their own offspring). These new frameworks have given rise to moral rights and obligations that weaken the old patriarchal and absolutist structures of past traditional families. For example, grandparents have gained certain legal rights over their grandchildren in the United States, and single parenthood has become overwhelmingly accepted in many countries throughout the world. Another change in the traditional relations within families results from new technologies such as IVF (in vitro fertilization) sperm donation and DNA testing. These technological innovations enable widows to bear children from their deceased husband' sperm, mothers to bear children from their deceased

---

[3] Habermas Jurgen. *The Future of Human Nature.* Polity Press. 2003. p. 42.
[4] Gordon Graham. *Genes.* Routledge. London and New York. 2002. p. 159.

son's sperm, or simply allow infertile couples to bear children from sperm or egg donations of strangers.

Thus, this section is devoted to some of the issues that have arisen due to these changes, such as the rights of adults born as a result of donated insemination and the rights of grandparents and other extended family members vis-à-vis the parents. Our attempts to navigate between the conflicting claims of different parties of the extended family sometimes raise very complicated issues, and these issues are examined here.

Chapter 8: Rights of Relatives and Generations

The presupposition of dealings in rights of family matters has always been that parents are the ultimate authority in anything having to do with (their own) children. However, the waning of traditional norms brings about challenges to this assumption. In this chapter we address the rights of relatives beyond the nuclear family. The first issue discussed in this chapter are grandparents who insist on their rights in seeing, meeting, or maintaining contact and relationships with their grandchildren against the will and preferences of the parents (who are the grandparents' children). The issue of grandparents' rights has become accepted in US Courts as all 50 states have laws that acknowledge such rights, at least to some extent. My view regarding this issue acknowledges and respects these rights and includes them within the large framework of rights within the family. In my opinion, though grandparental rights are subordinate to parental rights, they should still exert considerable weight when there is a conflict. Of course, the decisive consideration in resolving the conflict of parental rights with rights of grandparents should remain the children's welfare.

The second issue of this chapter relates to the controversial case in year 2000 of the Cuban child rescued from drowning and brought to the US where his mother's relatives claimed custody over him against the claim of the Cuban father (the mother drowned in the attempt). Although the boy was rightly returned to his father in Cuba, this case exposed some ugly aspects of American society though ultimately, the superior status of parental rights over the rights of other extended family members was maintained.

The third issue discussed in this chapter is a new law in Britain which allows children who will be born in the future of sperm-donations to reveal the identity of their biological fathers, who will no longer have the right to anonymity. Apparently this decision equates the rights of children who were born from sperm or egg donations to those of adopted children, who can see the adoption files when they turn 18 and become adults. (This will affect only future donors and children, and not those in the past whose anonymity will still be maintained.) In this chapter I discuss how such a law changes sperm and egg donations within the concept of biological parenthood.

Chapter 9: Procreation After Death

A more complicated issue that relates to parenthood and family is the desire of parents or spouses who have lost loved ones to use the deceased's sperm in order to create a new generation of that same family. There are a number of subtle moral dilemmas: often, it is not technically possible to obtain informed consent of a dying man. Also, a child produced from the sperm of a deceased man, will be the biological grandchild of the man's parents who will raise him as their child, causing a skip-over or a confusion of generations. Finally, there is the dilemma of bringing a child into the world as a 'monument' to a deceased family member.

This chapter deals with two different requests, each with its own complexity.

The first case is the request of a young widow to use the sperm of her deceased husband to have his child. Although the sperm was harvested from the husband before his tragic and unexpected death, he was not able to give informed consent as he was unconscious. However, his parents and the widow's parents all gave their consent and agreement to support the young widow in raising the man's biological child.

The second case is the request of parents to use the sperm of their dying son to create another child; in this case, the dying son was able to give his consent. The moral issue here is of a child born to much older parents who might not be able to take care of him appropriately over the long term, and might even need the new child to take care of them due to their advanced age.

Chapter 10: Babies as Commodities

The dimensions of global trade between rich and poor countries—certainly a direct consequence of current globalization—has both transcended traditional state boundaries as well as transformed anything and everything into objects of trade and commerce. That human beings have been used as commodities is familiar in our history: witness slavery and prostitution. However, the extension of this regrettable phenomenon to babies is one of the most objectionable aspects of our new global form of life and thought. What started as a generous movement of international adoptions, with well-meaning motivation and intention, has deteriorated into a capitalistic profit-making venture in which babies are no more than a means of maximizing profits. In this chapter we endeavor to pose, and answer, queries about the ethical implications of the tragic move from adoption to baby-commerce.

The specific case is a story about an adoption agency in San Diego that first offered twin girls for adoption to a couple from California and then took the babies back shortly afterwards to offer them to a couple from Britain, who evidently paid more than the first couple. The whole story was exposed after the babies arrived at the new home in Britain. This story ended after 4 years, when the twins were finally restored to the custody of their biological mother. But the entire case raises the specter that children are becoming just another form of merchandise for sale in the Internet, as "paid adoption" may cross the line into actual baby commerce. This chapter discusses the meaning of commodification of human beings in general, and of children in particular.

SECTION D. PUNISHMENT

Several decades of theoretical discussion concerning the concept and social function of punishment within society have yielded a number of important principles, leading to a reasonable understanding of punishment. These principles are explicated in this section, and include: the proportionality principle (in which the punishment must fit the crime), elements of deterrence (as in the consequential or utilitarian approaches), and humane treatment of convicts. However, several marginal and problematic situations still remain where additional considerations must be entertained, situations that vex the ethicist bent on making theory adhere to praxis. One acute problem in the current practice of punishment is that it is imposed by imperfect legal systems, whose obligation is first and foremost toward the rule of law, and not necessarily to moral or ethical considerations. This may cause distortions in the implementation of punishment, such as overly lenient penalties (in domestic abuse, for example), and exceedingly severe ones (such as the death penalty for the mentally retarded). I examine the complexity of these issues in the different cases I bring up in this section.

Chapter 11: Punishment of Sex Offenders

This chapter first discusses general characteristics of punishment as a whole as the background for the expectations we have for the punishment of sex offenders. These characteristics include the proportionality principle in which the punishment must fit the crime. However, I take into consideration that our usual guidelines regarding punishment are somewhat insufficient when we discuss punishment for sex crimes, due to the propensity of these types of crimes to arouse extreme emotional reactions.

I discuss two aspects of punishing sex offenders. The first deals with the outrageously mild punishments that are meted out to sex criminals in Israel, both in length of incarceration and in severity of punishment (sometimes minimized to community service), particularly to teenage sex offenders. This strikes at our deepest instincts regarding the proportionality principle in which the punishment must fit the crime. It also begs the question whether the tacit acceptance of society of these mild punishments, indicates that society accepts these values and norms behind the punishments involved. Do most people agree that men are superior to women, and that violence is a legitimate means for achieving men's goals and desires? If society itself does condemn such offences, then it follows that the courts that represent this society, should impose much more severe penalties for sex crimes.

The second aspect deals with the offering of plea bargains by the prosecution, in which curative treatment to the sex offender is a stipulation for reduced punishment. It is a relatively new idea of mitigated punishment pending the criminal's consent to chemical castration. In this case I argue that rehabilitation is not part of the prosecution's role as a legal authority.

Chapter 12: Punishment and Domestic Violence

Domestic violence is not just a sub-category of violence in general. It is inherently complex because there is an unclear border or thin line between punishment (disciplining children) and actual violence (beating children). Another problem is that when the offender is a parent, for example, then punishing him or her is likely to harm the family at large, such as by taking away their means of support. These complex issues are often used by the authorities as an excuse to abstain from pursuing and punishing offenders. Consequently, they renege on their two obligations—to punish criminals and protect the weak. I present the argument in this chapter that the privileges of autonomy and non-interference generally accorded to the family by the State, are privileges that are conditional on the proper functioning of the family. When domestic violence rears its head, these privileges of autonomy are cancelled and the family is, indeed, subject to State control and interference.

The second problem, more specific to Israel, is the intolerable laxity with which domestic violence is treated by institutions in Israel, most notably by the legal and law enforcement systems. In general, the legal system in Israel exhibits outrageous clemency toward criminals in general, and for domestic violence in specific.

Chapter 13: Capital Punishment and the Mentally Retarded

In this chapter I focus not on the general debate regarding capital punishment, but on applying the death sentence to mentally retarded or mentally ill criminals. One of the most relevant terms for discussing the legal and moral accountability of the mentally retarded is that of *mens rea* (intent required to commit the crime). I argue that this notion of diminished responsibility should preclude the imposition of capital punishment on mentally retarded and mentally ill people, without entering into the ideological dispute regarding capital punishment per se.

Examples are given to illustrate the troublesome phenomenon of Texas courts that are not much swayed by the impaired mental condition of offenders when deciding to impose capital punishment on them. I conclude that even those who support capital punishment should demand that it be imposed not only on those who deserve the most severe penalty allowed by society, but also those with the highest level of *mens rea*. Thus when trying mentally retarded or insane criminals, their mental state should be considered as a mitigating factor in reducing the death penalty to life imprisonment.

SECTION A. INDIVIDUAL RIGHTS AND PUBLIC DUTIES

This section deals with the inherent tension between the individual and the community. A person is both an individual as well as a citizen or a member of community, and this immanent dichotomy always involves conflicts.

These conflicts may occur between individuals who have their own preferences, desires, needs and interests, and the social framework within which the individual exists, which also has its own interests. The core of this conflict is the fact that many interests of the individual are acknowledged by society as valid claims against the social framework, and thus receive the status of *rights*. On the other hand, the individual, as a member of society, has obligations to fulfill toward the social framework, and thus is considered as having *duties*. Sometimes, the rights (of society) and the duties (of the individual) conflict with each other, and we have to decide which interest is more urgent and, therefore, should prevail.

Whenever there is a conflict between the individual's interests and those of the social framework, my own tendency is to support the individual's interests and provide them with more weight than the interests of the collective framework; this arises from a belief that the state or society is innately robust and can protect its interests more efficiently than the individual. However, this preference is not absolute and when the interests of the individual are only slightly harmed while the social framework is significantly benefited, then I accept the price of a minor violation of the individual's rights—but not without reservations.

This section deals with a few different domains within which such conflicts sometimes occur. The first chapter deals with the problem of privacy and the public sphere from different points of view. It questions the immunity from, or privilege against, disclosure of private information accorded to the individual—public figures, ordinary people, and even criminals—whereby it conflicts with the public interest. In this regard we deal with the question of medical confidentiality of public figures or elected representatives; exposure of the criminal records of minors who have grown up and want to be integrated into society; disclosure of names of public figures being investigated in criminal procedures, and so on. What I stress in this chapter is that there is a big difference between an ordinary person's claim for respecting his or her privacy, and the same claim when it is demanded by a public figure. I show how the issue of privacy becomes very complicated when it clashes with the public interest.

The second chapter deals with a more general topic: the obligation of the state toward individuals. Here we deal with the role of the state in protecting the welfare and personal security of deprived sections of society (including both citizens and residents). By "deprived" we mean children (or fetuses), mentally retarded individuals, or persons on the margins of society (minorities, fringe groups, etc.). We ask whether health and educational authorities are permitted to coerce individuals into normative forms of behavior, and whether they may desist from supplying services or basic necessities to such individuals. In this context we raise the question of the parental rights of mentally retarded people and the state's duties to them and their children. In this chapter, I try to express an instrumental concept of the state and argue that the state is obligated to provide necessary means to its residents, even in cases where the original duty to provide such means was imposed on others (their parents, for example).

The third chapter, "Public Security vs. the Right to 'Be Let Alone,'" deals, again, with the problem of privacy, but from a different perspective: that of databases. This has become a burning issue since September 11, i.e. since "terrorism" made its way into public discourse as the most intimidating phenomenon to date. Thus for reasons of (purported) security, the state inquires into such issues as the question of the legitimacy of DNA data resources that may help to locate or identify criminals, and the question of trace and documentation of telephone, e-mail, and other means of communication. Our own statement, in light of the current political ambience, is that there is room for worry that authorities might abuse the populace's concern about security and violate basic rights in so doing. Here, as in the first chapter, I tend to accept the establishment of a very specific database, but with stringent restrictions on its use. When there is a demand for a comprehensive database, I outright reject it.

The fourth chapter deals with freedom of expression in academia and the media. This chapter turns to different aspects of freedom of expression, which have characterized the ongoing debate in both academia and media. We ask about the academic freedom to express controversial views; for example, views that are couched in anti-feminist or anti-minority (call them racist) utterances in the classroom, on campus, in newspapers, on TV talk-shows, etc. A related issue, also addressed here, is the right (or duty) of the media to broadcast offensive or extremely gory events and images on television. We inquire whether it is either prohibited or compulsory to bring such images to public awareness. Here, too, even though I try to maximize freedom of speech, and allow "hate speech" at least in academia, I do not think that everything is worthy of being visually revealed to the public. I basically object to external censorship in the media, but in certain cases I expect the media to conduct self-censorship—not in verbal reporting but in televising brutal or gory scenarios.

Chapter 1

PRIVACY AND THE PUBLIC SPHERE

The right to privacy has become one of the most fundamental rights of the individual within society, and it has been widely acknowledged—both among legal systems in liberal states and among theorists of rights—ever since 1890, when Louis Brandeis and Samuel Warren wrote their famous article "The Right to Privacy."[1] There, they laid down "one's right to be let alone" as an essential part of one's autonomy, and insisted that one's private life, habits, acts, relations, etc. should be respected and immunized against external invasion or interference. We think of this right as having a special role in modern life and assume that, as Greg Pence puts it, "some rights of non-interference and some liberties are necessary to the minimally smooth functioning of modern society as we know it."[2] Since 1890, it has increasingly become an important task for Human Rights activists and organizations to protect this right of the individual both from governmental invasion and (even more) from the curiosity of the press and the public.

However, like every other right, the right to privacy can sometimes be used (or may we say "misused" or "abused") for harmful or immoral purposes. In this chapter, I examine three different cases where an insistence upon the respect for an individual's right to be let alone may cause some distortions in the original positive intention which led to the establishment of this right, and may bring about undesirable consequences. I want to show that even though we should be careful with possible invasions into one's private sphere, we have to make sure not to *overprotect* one's privacy, and use this right as an excuse for malicious intentions.

---

[1] Brandeis Louis and Warren Samuel. "The Right to Privacy." In: *Harvard Law Review*, Vol. 4, No. 5, December 15, 1890.

[2] Pence Greg. "Virtue Theory." In: Singer Peter (ed.), *A Companion to Ethics*. Blackwell, Basil, 1993. p. 255.

The cases I deal with exemplify how values, rights, and liberties that have been gained after long struggles and sacrifices as a means for the emancipation of the individual, can become instruments for more oppression by the authorities.

CASE 1

The first example which I believe demonstrates the possibility of negative use of the right to privacy, is the case of the frequent demand to ban the public exposure of the identity of people who are under police arrest or criminal investigation, or who are suspected of being involved with crimes. Usually, it is the pressure of public figures, members of parliaments and other celebrities that supports this demand for privacy. In addition to the healthy instinct which leads the ordinary citizen to object to such demands for privacy when they are made in support of the famous and powerful members of society, there is also a sound and critical reason for freedom and democracy seekers to mistrust the intentions of those who want to hide the identities of people who are under suspicion of wrongdoing.

The main argument of those who want to promote this initiative for the preservation of secrecy is that the disclosure of the identity of a suspect may ruin his or her "good reputation," and this itself may have negative effects on the ongoing investigation. They support their demand for privacy by relying on everyone's right to privacy and confidentiality, which they assume to be something which overrides the "public's right to know," the public's morbid interest in such information and the freedom of press. In order to have a better understanding of this claim for privacy, particularly when it is claimed by certain groups within a specific context, we have to question the foundations of such a claim. The real danger and the possible harm involved in the loss of a good reputation will be caused to those suspects who will not be prosecuted at all at the end of the investigation. If they will be prosecuted, the principle of the publicity of the law and the demand for an explicit and due legal process, will lead to the disclosure of their identity in any case. In both cases (i.e. whether charged or not), these suspects have the right not to be harmed more than the due legal process requires. The real question here is whether or not this right overrides the public's right to know, which is so vital for the public's protection from the government's arbitrariness and tyranny, and the public's main instrument to defend itself against governmental corruption and bribery. After all, we have learnt from Thomas Nagel that "Freedom of the press and of public dissent protects everyone against abuse of power and official harm and neglect of all kinds."[3]

When we speak of ordinary people who are suspects, there is still formally a public interest in the case (since in the criminal law the prosecutor is still the state, as representative of its citizens), but, in fact, the general public is not particularly interested in the specific identity of the person who is suspected of committing

---

[3] Nagel Thomas. *Concealment and Exposure*. Oxford University Press, 2002, p. 42.

a certain offense. In such a case the question of the disclosure of their identity may be negotiable and we may have doubts whether or not such disclosure is necessary or whether the issue of privacy overrides the public's right to know. In such cases we may well have considerations about the weight we should ascribe to the right to privacy and to the public's right to know. The main consideration here is what kind of offense is involved. In violent offenses, even a suspicion is sufficient for overriding the right to privacy of the alleged offender. In the event that the suspect is a non-adult or underage, his/her right to privacy overrides the public right to know (and this immunity is relevant even after the prosecution, until the offender reaches adulthood). In any event, the right to privacy of ordinary people can indeed, sometimes, override the public right to know, depending, of course, on the kind of offense that is involved and the age of the offender.

However, the abuse of the right to privacy is more probable when we deal with public figures that are likely to wield more influence on the legal and investigating authorities than ordinary people can. For example, when a high-profile person is being investigated and this fact is kept secret, this person can use his/her power or influence on the prosecution or the police to abstain from prosecuting or bringing his/her case to the court. A good example of the dangers of such secrecy was the investigation against the Israeli Prime Minister, Ariel Sharon, which was kept secret during the 2002 election campaign, and those who intended to vote could not know that they voted for or against a candidate who was suspected of committing offenses. The ability of ordinary people to use their immunities against the disclosure of their identities in order to gain profit from the public's ignorance about the investigation is much smaller than that of prominent or very wealthy personages. Of course, the latter can argue that the possible damage to his reputation or image is much greater than that of the ordinary person, so the risk involved with the disclosure of the fact that he/she is under investigation, should provide him/her with greater immunity against such a disclosure.

The fear that high-profile figures may abuse the right to privacy has established a norm in which "public figures" do not posses such a right (at least with regard to their habits, behavior, social connections, etc.). A public figure has been defined by William Prosser as a person "who, by his accomplishments, fame, or mode of living, or by adopting a profession or calling which gives the public a legitimate interest in his doing, his affairs, and his character, has become a 'public personage.'"[4] Another definition for "a public figure" we can take from Richard Hixson, who considers such a figure as "a person who, by virtue of a public exposure, generates an interest in her or his life and likeness."[5] Under this vague definition of "a public figure" we include not only elected figures or representatives, but also every person who may have influence on public opinion (celebrities, artists, people from the media, professional sportsmen, etc.). The rationale behind the norm

---

[4] Prosser William. "Privacy [A legal analysis]." In: Schoeman Ferdinand D. (ed.), *Philosophical Dimensions of Privacy.* Cambridge University Press. 1984. pp. 118–119.
[5] Hixson Richard F. *Privacy in a Public Society.* Oxford University Press. Oxford. 1987. p. 141.

of the shrinking of public figures' right to privacy to the bare minimum, is that since such people influence the public more than others, the public should know as much as possible about them in order to be able to assess and make reflections on their influence.[6] This may reduce their ability to abuse their influence on the opinions of the public. Such a rationale was the justification used by those who disclosed the identity of those married public figures who demanded the dismissal of the US president William Jefferson Clinton because he was involved in affair with an employee in the White House, while they themselves were involved in extramarital affairs. Those who disclosed the names argued that such congressmen and senate members are hypocrites, and saw it as their mission to warn the public of the hypocrisy of these representatives. When one speaks of moral values and principles, it is rightly assumed that critics of this sort must follow and respect these values and principles themselves before being permitted to censure others about their behavior. However, for those who are not considered as public figures, the right to privacy in such matters is strictly respected in our secular and more latitudinary society. But this is not the case when others are suspected of being involved in criminal offenses.

If we still wish to strike a balance between the public interest (which is embodied in the public's right to know) with freedom of the press and the individual's right to privacy in such matters of private morality, we should insist on the distinction between public figures and ordinary people. If we still want to insist on the importance of a person's right to privacy, we may ban the disclosure of the identity of those under suspicion who are not public figures (excluding violent crimes). However, public figures should not have this privilege and their names should be disclosed whenever the police investigate them. They cannot complain against this "discrimination" because when they became public figures they entered the public sphere and hence must be ready to be exposed to public criticism and control to a much greater extent than in the case of ordinary people. So long as they play any role in the public sphere they should not enjoy the privilege of total anonymity, even if ordinary people may enjoy this privilege.

CASE 2

A more specific dilemma, which is connected to the right to privacy of public figures, relates to medical secrecy. Here, even though there is a difference between ordinary people and public figures, there is also a distinction to be made between

---

[6] Two additional reasons for the loss of the right to privacy by public figures are mentioned by William Prosser: "that they have sought publicity and consented to it, and so cannot complain of it; (and) that their personalities and their affairs already have become public, and can no longer be regarded as their own private business." See, Prosser William. "Privacy [A legal analysis]." In: Schoeman Ferdinand D. (ed.), *Philosophical Dimensions of Privacy.* Cambridge University Press, 1984, p. 119.

celebrities, artists, sportsmen, etc. on the one hand, and elected representatives and leaders of states and public servants, on the other. No doubt every person has the right to medical secrecy and there is no question that this is a significant part of one's right to privacy. In matters of this sort we should respect the right to privacy even when we deal with people who have influence on public opinion since their health is still private and intimate information, which they are surely entitled to keep secret.

One good example of this is the tragic death of the well-known Israeli singer, Ofra Hazza. Although Ofra died from AIDS, the details of her illness were publicized only after her death. Although she was very popular and famous, the media respected her privacy. She was neither an elected official nor did she hold a position of special responsibility and since her family expressly insisted on avoiding publicity regarding the nature of her illness, the media rightly considered the public interest in her illness, though understandable, as mere gossip. Thus the media rightly concluded that there were not sufficient grounds during Ofra's lifetime for overriding her right to privacy and medical secrecy.

This should be contrasted with another case that concerned a rumor about one of the candidates in the primary elections to the US presidency 2000 elections. The rumor was that this candidate suffered from a mental illness that was caused by an event that occurred during his military service. Since this impairment was rumored to possibly reflect on his abilities to function as a United States president, the candidate immediately made public all the details of his medical records including very private and intimate details, such as a scar on one of his concealed limbs. The "over exposure" of the medical record raised vociferous criticism of this candidate, but he responded that the voters had a right to be fully informed about a candidate for whom they are asked to vote. Hence they had the entire right to know every relevant detail about his health, assuming that his health was a relevant detail of his governing competence and capability.

Two other cases, this time relating to serving prime ministers, may serve to clarify the issue of medical secrecy of leaders of states. In the early 1980s, the Israeli Prime Minister Menahem Begin suffered from a certain disease that disabled him from carrying out his public duties as prime minister. The press refrained from reporting anything about Begin's health and only let it be known that he had secluded himself in his apartment. It took several months until Begin resigned and during all that time, the State of Israel functioned with an incompetent prime minister though Israeli citizens were not aware of this. When the matter became known it was widely felt that the Israeli press had acted irresponsibly, and violated the public right to know about their prime minister's ill health that directly impinged on his ability to fulfill his office.

The second case was in Japan in December 2000, when the Japanese Prime Minister totally lost his consciousness. Hours passed until the minister who had to take over was informed, and a full day passed until the public was informed about the whole story. Moreover, the government issued a vague and misleading official announcement and this, in turn, caused the newspapers to publish false information: the media reported that the prime minister was in full control of affairs, even though in fact he was unconscious at the time. When the truth was discovered, the humiliated press attacked the government for misleading the public and failing in their duty. In so doing they had placed the press in a situation where they were unwittingly part of a conspiracy plotted by the ministers of the government. The leaders of the government replied to this accusation by saying that they did not report the prime minister's true health condition only because of respect for his dignity and to honor his rights to medical secrecy. They thus totally missed the point made by James Nickel, that in such a case respecting and implementing the prime minister's right to privacy may "carve important exceptions into the right of freedom of the press,"[7] and the cost of their decision is much higher than a free press can tolerate.

It would be very easy for us to condemn the government, in this case, and to justify the press anger about it, and particularly the mistrust they feel toward the government. However, beyond the misuse and abuse of the right to medical secrecy, and the deliberate misleading of the population, the government's behavior may be understood as a reflection of a perverse political culture in which governing and even sovereignty belong to the established leaders who are in power, and the public and press should not interfere with the business of government. Such a culture could explain the attitude of screening and distorting significant information, even when this is private and intimate, so long as this is "for the good of the people." But to leave the judgment of what is "for the good of the people" exclusively with the existing wielders of government power, may lead to unrestrained governmental corruption. We could easily conclude from the government's conduct in this case that those who hid significant or sensitive information from the public did so for their own interests and reasons, and this may lead us to suspect that they had something to hide concerning their own interests and conduct. We can be doubtful about the reasons for hiding information from the public and can be cynical about their declared respect for someone else's privacy. In this case, the public's "right to know" seems crucial in order to protect the individuals from governmental arbitrariness and quite certainly overrides the right to medical secrecy of elected figures, particularly if these figures occupy the highest political positions and bear a great responsibility.

However, there is something else that we can learn from the above examples regarding the different norms relating to politics and the press in different countries. In the US, where people are more aware and worried about the freedom of the press and information, the government and the candidates are much better supervised by the press and the public. Hence the possibility of revealing corruption is higher and the civilians are less tolerant about it. In Israel, the norms of disinformation and screening are so rooted in the public sphere and political life,

---

[7] Nickel James W. *Making Sense of Human Rights*. University of California Press, 1987, p. 121.

Privacy and the public sphere 11

that sometimes the whitewash and cover up of scandals or corruption has come to be considered as a service to the nation or even patriotism. Frequently, the government hides information behind the screen of "national security" because of the respect it feels it owes to highly regarded functionaries, and many times even without any explanation or excuse. In the better case this may lead to the sacrifice of a single victim, but in the worse case it may sacrifice the future of the nation. If the press will not come to its senses and provide the citizens with all the relevant information about their leaders (including their ability to function), it may bear the responsibility not only for future damage caused by the hiding of information, but also for the scandals that were rendered possible due to this hiding of information. We should remember Denis McQuail's suggestion, that the media "have an implicit obligation to serve the public interest, by protecting and enlarging a 'sphere of the public' in matters of morality and belief as well as of information."[8]

These three examples relating to medical secrecy all demonstrate the importance of the public's right to know, and the danger that this right may be violated with the excuse of someone's right to privacy. This danger is acute when those who wish to abuse the right to know are governmental officials. The right to privacy, which originally intended to limit surveillance by government authorities,[9] is turned around by governmental officials to serve as an excuse for the authorities to escape the citizens' control. Since it is true, as Colin Mellors claimed, that "the possession of information is the possession of power,"[10] we should insist on explicit and open accessibility to the relevant and correct information about the competence and the abilities of our leaders.

CASE 3

The last example regarding the right to privacy is much more complicated. It is meant to show, again, how people may abuse the right to privacy and use it as an excuse to gain benefits to which they may not be entitled.

This time the example is taken from a decision made in Britain by the High Court Judge Elizabeth Butler-Sloss in January 2001. This decision granted two young murderers, Robert Thompson and Jon Venables, an open-ended injunction ordering that when they are freed their new identities must not be disclosed by the news media.

---

[8] McQuail Denis. "The Mass Media and Privacy." In: Young John B. (ed.), *Privacy*. John Wiley & Sons, 1978, p. 191.
[9] See, for example, Westin Alan. "The Origin of Modern Claims to Privacy." In: Schoeman Ferdinand D. (ed.), *Philosophical Dimensions of Privacy*. Cambridge University Press, 1984, p. 70.
[10] Mellors Colin. "Governments and the Individual—Their secrecy and His Privacy." In: Young John B. (ed.), *Privacy*. John Wiley & Sons, 1978, p. 109.

The two murderers were 18 years old at the time of the injunction. Eight years previously, in February 1993 when they were each 10 years old, they committed one of the most notorious and cruel crimes in Britain. They lured 2-year-old James Bulger to a place where they bit and stoned him to death, then left him on the tracks to be cut in two by a train. A jury found them guilty, assuming that at the time they murdered, they were competent to distinguish between good and evil. In 1999, Lord Woolf, the Lord Chief Justice of England and Wales, decided that they had been sufficiently punished and since they expressed regret, they could be released on parole in 2001, 8 years after the murder.

In her decision, Judge Butler-Sloss said that the teenagers were "uniquely notorious" and would be in danger if anyone found out where they lived. She said that they were at serious risk of attacks from members of the public as well as from relatives and friends of the murdered child. She granted them a privilege that is usually reserved for state's witnesses, and banned the publication of any information about their new living area, profession, appearance, etc. This decision aroused fury both among the members of the victim's family who thought that the sentence was so minimal as to undermine confidence in the UK criminal justice system, and also among the seekers of freedom of the press and of information, who claimed that this decision struck at their right to report and publish significant information about issues that concern the public interest. While the victim's family was concerned about the readiness of the courts to come to the relief of very young offenders and bestow unreasonably considerate treatment on cruel and malicious murderers, the seekers of freedom of information were concerned about the possibility that other criminals might enjoy the same immunity in the future.

The rationale behind Judge Butler's decision was the need to protect the murderers and make their rehabilitation and re-integration in society possible. However, when asking about a possible reasonable justification for this, the answer may be found in the young age of the perpetrators when they committed the crime. This sounds sensible since adult murderers are not protected in a similar way. It makes sense that the conjunction of the dreadful crime, the relatively light punishment they would eventually pay for it, and the fact that the public's (and family's) anger was greater than usual, all contributed to a situation which caused the two boys to be exposed to more danger than is faced by other murderers after their release. (The victim's father said that if he would run into them on the street he would, indeed, take vengeance). These circumstances made it necessary to protect them more than other released criminals. In any event their young age when they committed the crime was the main reason for the privilege they received from Judge Butler.

However, the decision to provide them with the same protection that is usually bestowed upon criminals who supported the law—such as state's witnesses—is indeed one that engenders rage and raises feelings of injustice. It might seem that the murderers enjoy more considerate treatment from the legal and social authorities than is afforded to the victim's family and the public. Even our awareness of the young age of the two criminals, and the understanding that they should be given considerate treatment, do not relieve these feelings of rage and injustice. I will try to suggest a possible rationale for these feelings and recommend a different way of dealing with the whole problem.

A possible reason for the indignation about the decision to grant them immunity from disclosure is the assumption that they already enjoyed considerate treatment by being released after only 8 years of incarceration, which is a very mild punishment for such a brutal murder. It looks as if the murderers are being given the same consideration twice: once in being given a light sentence and again by given the privilege of concealment after release. If they were adults when they committed the same crime, they would not have been released so quickly nor would they have enjoyed any immunity at all regarding their identity after their release. The fact that they were so young has helped them already once, and now they have to face a new reality, this time as adults who bear the stigma—the "Sign of Cain" on their forehead. As an answer to the argument that says that they deserve a second opportunity in life, stands the claim that their early release while still very young already gives them a second opportunity.

The screening of information from the public is not only inefficient (since information cannot be censored on the internet, and since immunity is in force only as long as they continue to live within Great Britain), but also strikes brutally at the freedom of press and the right of the public (for example, the new neighbors and future employers of the two murderers) to know with whom they live and interact. It would not be right or just to treat them—due to ignorance or unawareness—as if they were innocent, or as if they were ordinary men, since they are not ordinary men. The crime they committed is part of their biography, and they will have to live with it. This crime is inexcusable and unpardoned, and even if their society allows them to keep on living freely within it, they still need to pay at least some price for what they have done and for who they are.

With regard to the claim that their lives are threatened by the possible fury of the public, and hence they should be protected and hidden, we can respond in two ways. One is that they still have the option of remaining in prison till the danger is past—i.e. until the punishment they serve is considered appropriate and thus calm down the anger and the will of the family and the public to hurt them. Then, we will be able to say that they paid the price for their crime, and thus everyone should leave them alone. The second is that if the authorities' problem is to physically protect them, they can resolve this problem by either guarding them or sending them elsewhere, to a place where there is no such a fury toward them (such as another country, as is usual in the case of state's witnesses). The solution of providing them with such immunity, which strikes at the freedom of press and the public right to know, is not the right one in this case. The authorities should search for a better solution that will not abuse the right to privacy and undermine the public right to know.

The conclusion we can draw from the above three examples is that even though the right to privacy is very important in modern life, we should be careful of its manipulative use. Abusing the right to privacy or using it as an excuse for concealing evil intentions may well reduce its significance in our life. If we really believe that the possession of information is the possession of power, the better way of diluting governmental power, as suggested by Colin Mellors, is "the reduction of the secrecy that currently surrounds governments and their civil servants. In simple terms, the best safeguard is not that they know less about us, but that we know more about them; and that we are aware of what they know about us and how they use that information."[11]

---

[11] Ibid.

Chapter 2

THE OBLIGATION OF THE STATE TOWARD INDIVIDUALS

This chapter examines some of the very sensitive questions about the obligation of the state toward individuals within it, particularly the very weak and deprived sectors of society. With regard to these sectors there is always the fear of either overprotecting their rights on the one hand, or of arbitrary, invasive, and even offensive treatment by the state, on the other. The state, per the *Parens Patriae*[1] doctrine is supposed to guarantee not only the freedoms and negative rights of its residents, but also their positive rights. However, occasionally it appears that the desire to fulfill this duty and protect certain rights and freedom of a particular individual, may cause harm or violate some other rights. This chapter will deal with certain issues where the proper extent of the state's obligations (if there are any) is not very clear, and will try to suggest limits and restrictions to this obligation. The specific cases to be examined are those of fetuses, minors, and mentally retarded people: three vulnerable sectors in society. In each case the examined rights will be different, but they illustrate the complexity of the general question about the general obligation of the state toward individuals within it—particularly the weak and the vulnerable. At times the principle obligation of the state is to provide its residents with certain capabilities and skills, but there are instances where this obligation requires the state simply to abstain (ensuring that others will abstain as well) from any interference with one's liberty, and to respect the individual's freedom.

---

[1] *Parens Patriae* is the doctrine whereby the state takes jurisdiction over a minor living within its border. Usually it is the basis for deciding what state will assume jurisdiction in a child custody case, but it also used for ascribing responsibility and accountability on the state toward children within its borders.

CASE 1

The first case concerns a decision made by the Bristol County (Massachusetts) Juvenile Court Judge, Kenneth P. Nasif in 2000, to send to jail a 32-year-old woman, Rebecca Corneau, who was pregnant at that time. His intention was to incarcerate her in prison until the birth of the baby, then take the baby into custody and release its mother.

Mrs Corneau was a member of a cult from Attleboro, Massachusetts which does not believe in or cooperate with modern medicine. A year earlier she had given birth to a stillborn baby, and the police suspected that any competent physician present at the birth would probably have been able to save the baby. Since two other babies from that cult had died during medically unaided delivery, the Judge thought it is his duty to protect the Corneau's future baby as well as other children from that cult. Hence he denied custody to the Corneau family and to two other families of the same cult, and appointed guardians to their existing children. Bristol County District Attorney, Paul F. Walsh, said that though he had doubts about the legal grounds for the decision to arrest the woman, he felt confident on its moral grounds.

My purpose here is to examine the decision to arrest Mrs Corneau in order to protect the fetus. I do not want to enter too deeply into the question of custody, but to use this example for discussing the alleged obligation toward fetuses. I know that there are many difficulties even with the question about the parental competence of Mr and Mrs Corneau, but at this point I content myself with citing Edgar Page's position, which says that "we must remember that parental rights are those rights which people have simply *as* parents, not as *good* parents."[2]

The legal question here, which is marginal to my discussion, raises the question about the appropriateness of the Judge's decision. When even the District Attorney said that he had hesitations about the legal grounds for a decision that supported his side, the Judge had to rule against the DA, even if only due to doubts or insufficient evidence, since he did not have certain and unquestionable legal grounds for a decision against Mrs Corneau. However, what is more interesting and important here is the question of why the District Attorney was so confident about the moral and ethical grounds for that decision. With regard to this question we can inquire into three main issues.

The first relates to a situation where one's freedom is violated not only without committing an offense, but without even being suspected of committing one (Mrs Corneau's husband was incarcerated for 138 days for not reporting what happened in the previous birth). One can interpret the situation as the incarceration of an innocent person, and therefore cannot be justified. The freedom against arbitrary

---

[2] Page Edgar. "Parental Rights," In: Almond Brenda and Hill Donald (eds.), *Applied Philosophy: Moral and Metaphysics in Contemporary Debate*. Routledge, London and New York, 1991, p. 75. The emphasis in the original.

arrest is so fundamental that its violation requires substantial and solid grounds. Future possible harm that may be caused to the baby cannot be sufficient for arbitrary arrest.

The second issue relates to the moral status of the fetus. Even legal commentators, and certainly ethicists, think of this issue as open to dispute. To be sure, there are commentators who think that a fetus has moral standing, and hence is entitled to possess moral rights. Joel Feinberg, for example, believes that "unborn children are among the sort of beings of whom possession of rights can meaningfully be predicated."[3] According to such a view, the Judge's decision to nominate an advocate for the fetus was correct. However, it is quite a stretch from this claim to the claim that an uncertain interest (since it is not clear that during or after birth the baby will be exposed to a threat or danger) of a fetus (which is a future moral subject), overrides the interests of the mother (which is an actual moral subject) to be protected against arbitrary arrest. When there is a conflict between an interest of a future moral entity (the embryo), and an interest of an actual moral entity (the mother), the interest of the mother should override that of the embryo, primarily since she can make explicit and strict claims about her interests, freedom and rights.

Another indication for the priority of the mother's rights is the legitimacy given by many theorists to terminate a pregnancy if the mother's life or health is in danger, even among those theorists who object to abortion. One example is Richard Werner, who holds that abortion is morally allowed only during the first 8–10 weeks of pregnancy. However, even he believes that abortion can be morally justified at a late stage of pregnancy "whenever the mother's life was endangered by a continued pregnancy."[4] The position, which claims that the state has to protect unborn babies against the interests of their parents, has already caused many evils. Judges have punished pregnant women who were convicted for minor crimes for longer periods than they deserved, in order to make sure that they would give birth to their babies inside prison. One of the reasons for doing this was to take the newborn babies away from their mothers and take them into custody. By doing this, those judges had grossly violated *Lex Talionis* (the law of retaliation)—a principle that demands proportionality between gravity of an offense and severity of its punishment. A woman's rights to due process and to equality before the law are still valid when she is pregnant, especially when there is no conclusive evident that the unborn baby is or will be in any danger. Thus, we can conclude that an abstract interest of the fetus cannot override the mother's right to due process, and this conclusion should apply also to the case of Mrs Corneau.

---

[3] Feinberg Joel. *Rights, Justice and the Bounds of Liberty: Essays in Social Philosophy*. Princeton University Press, Princeton, New Jersey, 1980, p. 180.

[4] Werner Richard. "Abortion: The Ontological and Moral Status of the Unborn." In: Wasserstrom, Richard A. *Today's Moral Problems*, Second Edition. Macmillan Publishing Co. New York, 1979, p. 73.

The third issue regards society's attitude toward those who do not adhere to the prevailing consensus or belong to the majority group. Even if the woman's faith, beliefs, or lifestyle does not accord to those of the judge or the mainstream of her society, she still deserves the same protection from governmental or legal arbitrariness, stupidity or harshness that is afforded to every other member of her society. Her freedom, welfare, and faith should be protected with the same determination we protect the interests and values of others whose belief or faith *are* part of the prevailing consensus. As we do not deny custody to parents who believe that a certain prophet dramatically disappeared into the heavens (or will return from there in the future), we should not deny custody to parents who do not believe in modern medicine. If there is the slightest evidence that members of certain cults (or any other parents) intentionally harm their children, they should be put on trial and their children may be taken from them. However when there is not sufficient evidence for this, we cannot consider their different lifestyle or faith as a reasonable justification for the violation of some of their fundament rights. It is the members of minority or oppressed groups who need protection from the arbitrariness or despotism of the majority, not a fetus to be protected from its parents.

It can be claimed in this example that sometimes the state needs to protect future persons, by violating the freedom of an existing person. We can be skeptical about the sincerity of the authorities in making this claim, and suspect that the authorities are far more worried about the cult that Mrs Corneau belongs to, than about the chance that she (or her husband) will hurt their children in the future. To sum up: we argue that in this case the obligation of the state is merely to abstain from any action, unless it has evidence that someone is in actual danger. Until then, its obligation is to respect the parental rights of its residents. The next example may show us that the state is not as concerned about the future of its children as it would like us to think, and allows itself to be indifferent to the education they acquire, even from public institutions.

CASE 2

The second issue concerns the state's obligation toward its citizens as a result of the possible consequences of inappropriate education. I refer here to a specific example, but the general problem is more important than that specific case. I deal here with a case where a group of young men who studied in ultra Orthodox institutions decided to sue the state of Israel for its failure to provide them with the skills and education that are necessary for integration into modern society. They explained that the education received at these institutions is primarily in Jewish religious studies with a smattering of reading, writing, and a little arithmetic. There is no English, mathematics, physics, chemistry, geography, etc. in the curriculum. Consequently they had only unsatisfactory, low-income employment opportunities. They claimed that the institutions in which they studied operate under the compulsory education law and are financed by the government. Hence, it is the government's duty to enforce pedagogical supervision at these institutions. The state's negligence of its obligation, allows these institutions to abstain from teaching (or teaching at appropriate level) those subjects that are essential for most of the desirable and rewarding jobs in modern society. They are also crucial for entering institutions of higher education, where the necessary skills for attaining prestigious or desirable jobs are acquired. Such a selective (or shall we say "poor") education places the graduates of these institutions at the bottom of the socio-economic scale, with the inevitable dependence on charity, welfare, and political patronage.

In a similar suit, prosecuted in the US, a school graduate sued his school for not providing him with the necessary skills in reading and writing. He won the case and was awarded by a large sum of money. The judge in that case decided that it was the high school's mission to ensure that its pupils would realize their elementary right to learn how to read and write. However, the new element in the case I discuss here is the claim that the State should impart not only the ability to read or write to its residents, but other skills as well that are essential for social mobility and economic security. It is obvious that those who are not provided with these skills are sentenced to a low socio-economic status. What we should ask here is whether the state bears the duty to offer certain skills to its residents within its compulsory educational system, and whether the state is the only entity that bears this duty.

It appears that the state's obligation to provide a decent education, at least in the USA, was acknowledged many years ago by the Supreme Court even during the now-discredited era of "separate but equal" treatment of blacks and whites. In the case of *Sweatt v. Painter* (1950)[5] the Court ordered the School of Law in the University of Texas to admit an African-American young man, Herman Marion Sweatt, despite the ban on admitting African Americans to universities. The reason for this decision was the state's failure in providing equal education for all races, since there was no way through which African Americans could acquire legal education or to study law as there were no separate law schools for African Americans. A more significant decision was the case of *Brown v. Board of education* in 1954,[6] which virtually forced the admission of an African American to a school of white children. This time the reason was that children from minority groups studied in segregated schools that suffered from inferior

---

[5] *Sweatt v. Painter*, 339 US 629, 70 S. Ct. 848. Reference from: Dworkin Ronald. *Tanking Rights Seriously*, Harvard University Press, Cambridge, Massachusetts, 1977, p. 223, note 1. This case was argued on April 4 and, decided on June 5, 1950.

[6] *Brown v. Board of Education*. 347 US 483 (1954). Argued December 9, 1952. Reargued December 8, 1953, decided May 17, 1954.

resources and facilities, and did not provide an education equal to the schools of white children. The court decided that it is the state's duty to guarantee genuine equality of education between different races.

The second question, whether the state is the only liable and accountable body for providing decent education to its residents, and hence is the only relevant respondent of such a suit, is much more complicated. If such a suit were filed by people who lived in sparely populated areas without alternative schools that could provide a better education, the case would be similar to that of Brown. However, in the same towns and cities where these young people studied in ultra-Orthodox institutions, other public and state schools that provided better education and skills were, in fact, available. Thus, the state could claim that it was the parents' fault for sending their children to these very same schools, and not to other and better schools.

In my opinion, such an answer of the state is without basis, mainly because those young people were underage when they went to school. This fact imposes on the state the duty of protecting their rights and interests, and particularly their right to an open future—as being their *Parens Patriae* in every case where parental behavior may threaten this right. The state usually acknowledges this duty in many domains of life: the fact that it legislates a compulsory education law and requires that parents vaccinate their children, indicates that the state admits its status as guarantor for the fulfillment of children's rights, even when these children's rights conflict with the preference of their own parents. Therefore, the state can interfere with the parents' autonomy in the domain of education. Unlike the previous case about the parental rights of the cult members, where there cannot be a compromise between these rights and the state's desire to deny custody, here we can let the parents educate their children in a way that expresses their religious beliefs but demand that in addition they also teach general subjects. In this way the state's interference still respects their parental rights, but also protects the children's future interests.

The second claim, which the state could raise, is the autonomy it bestows upon the different educational system with regard to the curriculum. However, this claim cannot justify the repudiation of its duty to provide a decent and effective education to the graduates of these institutions, since, again, they were underage when they had studied at these schools. Ultra-Orthodox schools should still be allowed to teach more religious subjects and hours than other schools, but they should, at the same time, provide their pupils with general and elementary education. If the state was careless about its duty toward these young people with regard to their early education, it is now its duty to compensate them by providing them with appropriate professional training. The state should not desert them again and leave them without the necessary skills to conduct the modern economic struggle for survival that they were never prepared for.

The state's obligation in such cases cannot be summed up merely by providing free education, and in the conferring of autonomy upon the different educational

The obligation of the state toward individuals 21

systems. It has to interfere in the curriculum of the schools and make sure that what it considers crucial for guaranteeing the child's right to an open future will be included in the curriculum. Negligence concerning this duty entails that it will have to support the adults and compensate them later on in life by providing professional education that may allow them the possibility of decent future. The need for survival requires that one will have the necessary means for this (according to the Kantian principle "Ought Implies Can"), and professional skills are among the basic means for survival. If we really believe that these skills are indispensable and we follow Henry Shue's argument that "If everyone has a right to Y and the enjoyment of X is necessary for the enjoyment of Y, then everyone has a right to X,"[7] we must conclude that it is everyone's right to possess those skills, and it is the state's obligation to provide the necessary education which leads to the possession of these skills.

This concept of the state's role was discussed in detail by Joel Feinberg, who developed the idea of "the child's right to an open future." We can sum up this right as the right to "healthy well-rounded growth into full maturity as citizens with all that this implies in a democracy."[8] This is to say that a child has the right to be well-protected, well-educated, to be physically and mentally nurtured, and to acquire all the cognitive and mental abilities that are necessary for a decent and fair maturation. Usually this right is imposed upon the parents. However, Feinberg says, "The existence of such a right ... sets limits to the ways in which parents may rear their own children, and even imposes duties on the state, in its role as *parens patriae*, to enforce those limits."[9] This means that when the parents fail to rear their child properly, in a way that promotes his/her self-fulfillment, the state is supposed to restore this.

CASE 3

The third case I want to study in this chapter is a decision that was given by a British court, which prohibited a mother from castrating her 28-year-old son who suffered from mental retardation. Her main argument was that his castration would improve the quality of his life, since it would avoid the need to control his sexual life in the future. At that time she closely controlled his sexual life and this significantly restricted it. Her son was already involved in sexual intercourse with women, and he was probably capable of impregnating a woman. The court

---

[7] Shue Henry. *Basic Rights*. Princeton University Press, 1980, p. 32.
[8] This is a quotation that Feinberg brings from the US Supreme Court decision in 1944, in *Prince v. Massachusetts*. See Feinberg Joel. *Freedom and Fulfillment*. Princeton University Press, Princeton, New Jersey. 1992. p. 80.
[9] Feinberg Joel. *Freedom and Fulfillment*. Princeton University Press, Princeton, New Jersey, 1992, p. 89.

rejected her request and thought that the castration was not necessary so long as the mother was able control her son. In the event that he would be remanded to the authorities' custody, the court would reexamine the question whether such a castration would improve his freedom or quality of life.

The sore question about the right to procreation of mentally retarded people has been widely discussed during the last decades. The main argument of those who support the demand to sterilize mentally retarded people is that there is a high probability that the spouse will be mentally retarded, and that the offspring will be mentally retarded as well. Thus a baby born of such a union will need special treatment and the cost and burden of this treatment will be imposed on society as a whole, since the biological parents (or grandparents) will not be able to fulfill all his/her special needs and requirements. It is not very difficult to reject such an argument, since it tries to justify a physical offense on an individual on the grounds of social benefit. A person's right (even if mentally retarded) to bodily integrity always overrides social considerations, and we should not enforce an attack on a person's body in order to achieve external aims (or aims which are not that person's self interest). Any compromise concerning this strict principle may result in treating a mentally retarded person as an object. Such a compromise may lead to justification of restraining procreative rights to other, mentally healthy parents suffering from hardships such as poverty or insufficient education, on the grounds that the cost of the education or welfare of the children may also be imposed on society.

This approach toward the precedence of individual rights over collective goals follows Ronald Dworkin's "trump" theory. Dworkin thinks that "Individual rights are political trumps held by individuals. Individuals have rights when, for some reason, a collective goal is not a sufficient justification for denying them what they wish, as individuals, to have or to do, or not a sufficient justification for imposing some loss or injury upon them."[10] According to this view, the rights of individuals are more prevalent than those of society, and hence should have precedence over the latter.

A more serious and complicated argument is the one grounded in the concern for the welfare of a child who may be born imperfect, as a result of uncontrolled sexual intercourse between mentally retarded adults. Such a concern about this issue of the conflict between elementary rights of different subjects, requires a deeper and more serious consideration. This is because we believe that even a fundamental right of an individual can be overridden, but only when the right of another individual (or individuals) is more weighty than the overridden right in the situation under examination.

The question of the procreation rights of mentally retarded people is usually divided into two separate issues. The first is similar to the current example, and

---

[10] Dworkin Ronald, *Tanking Rights Seriously*, Harvard University Press, Cambridge, Massachusetts, 1977, p. xi.

deals with their right to give birth to children. The second deals with their right to raise their biological children. When someone tries to violate the first right under the excuse of the future welfare of the yet unborn baby, he/she makes illegitimate and even misleading use of the term "a possessor of a right." This is because we usually consider only *concrete entities* as "subjects" (or "possessors" or "holders") of rights, and hence we restrict the possession of rights only to "existing" or "actual" or "concrete" entities. Thus, the introducing of the right of a not-yet-existing entity (not even as an embryo) as conflicting with the right of the mentally retarded adult to bodily integrity is a fraud, since a non-existing entity cannot possess rights. Any attempt to strike at the right of the mentally retarded adult to bodily integrity under the excuse of protecting the right of a future entity, is only an attempt to present a cruel phenomenon as if it were enlightened. The current norms and conventions that are acknowledged as allowing intrusive medical treatment (specifically when an operation is involved), require that the one whose consent to medical treatment is denied will be "at substantial risk of serious physical harm or enduring suffering."[11] We also have to be sure that this treatment is for the patient's "best interests." However, while we do not have conclusive evidence that sterilization is in the mentally retarded person's best interests, we can be certain that this person does not suffer from any acute disease that allows us to undertake coercive and intrusive treatment. If we think of his mental deficiency as an acute disease it may justify medical treatment, but not castration.

A totally different issue is the second right mentioned above: the right of mentally retarded people to raise their biological children. When parents suffer from acute mental deficiencies, the physical and mental welfare of children in their custody may be threatened. Thus, while parents with mental retardation have full right to bring children into the world, they are often not able to care for these offspring appropriately. This creates a real conflict between the rights of the children to welfare and the rights of the parents to raise their children. Here two actual "subjects" of rights have claims, which can conflict in certain circumstances. The welfare rights of the children in such a case override even the fundamental right of the parents to keep and raise their biological progeny. In such a case, the rights of the children may justify removing them from their parents' custody and relocating them to a safer framework that will better protect and promote their interests. This is the price we have to pay if we allow mentally retarded people to have children. Here, a decision in favor of the children is possible, inter alia, since it does not include any surgical interference or any strike at the parents' right to bodily integrity. The question faced by the British court was different since it related to the request for physical harm to a mentally retarded innocent, made by his legal and acknowledged guardian. In the above

---

[11] See, for example, Campbell Tom and Heginbotham Chris (eds.), *Mental Illness: Prejudice, Discrimination and the Law.* Dartmouth. Aldershot, England. 1991. p. 124.

case, the court did not consider the guardian's consent as an appropriate substitute to the direct consent of the ward himself to medical treatment. A contrasting decision would have restricted the rights of mentally retarded people (which are significantly limited anyhow) to receive humane treatment from their society.

What I want to stress with regard to the second question—the rights of mentally retarded people to raise their biological offspring—is that in cases where the parents can do it safely (even with substantial support by the welfare authorities), the state has to allow them to do so and provide them with the necessary means. These resources may include financial aid, guidance, and even close and constant aid. In cases where the children's welfare and humane development can be secured within their parents' custody, there cannot be any justification for their removal from their parents' house. The only justification for this can be the absence of any possible way to leave them with their parents. The parents' rights are paramount, and should be carefully protected in these cases.

This example shows us that we have to be very careful and sensitive regarding the rights of mentally retarded people and minimize the restrictions on their rights. When we do determine that we must violate any of their rights, we have to be sure that it is due to the urgent and immediate interest of other subjects and not merely for reasons of social convenience or economy. In the current case, the state's obligation toward the individual means that it will protect his/her rights to bodily integrity and in case the parents need help in raising their children, the state's obligation is to provide them with any necessary assistance. In case those children need genuine protection it is the state's obligation to provide this, but the state should never exaggerate this need and violate the parent's rights without concrete grounds.

To conclude, we can say that the obligation of the state toward the individual imposes a great responsibility on the state, not only to respect and preserve the freedoms and negative rights of their residents but also to make sure that their positive rights will be fulfilled. This means that the state is obliged to provide a decent education, to support parents who have difficulties in raising their children, and sometimes even to protect children from their parents. However, the state should never use its obligation toward future or even actual residents as an excuse for violating the positive rights or freedoms of certain individuals. If the state does so, this may be justified only for the sake of other individuals' interests, not for the sake of the obscured interests of the state or any other collective framework.

Chapter 3

PUBLIC SECURITY VS. THE RIGHT TO "BE LET ALONE"

Issues of public or national security have spawned many questionable needs such as establishing new information banks and expanding existing databases that collect and retrieve private information about individuals. Such demands are usually based on two grounds: the need to fight criminality on the one hand, and the need to fight terrorism on the other. With regard to these needs, new ideas and suggestions arise on almost a daily basis. This chapter will examine two examples of such demands and reflect on possible implications and consequences of complying with them. The first example is the desire for a comprehensive collection of information about an individual, thus constituting a clear invasion of the private domain. I take the strict rejectionist position on this issue and bring several arguments to support my view. The second example concerns the limited need for access to specific personal information about individuals; in this case I argue for qualified compliance with the iron-clad proviso that close surveillance and control must be maintained over the access to such a database.

The idea of limiting or even violating certain rights including human rights due to reasons of national emergency is quite familiar. There are human rights theorists, and even human rights Conventions and Covenants, which permit the infringement of human rights during situations of national emergency. For example, James Nickel, a major scholar of human rights (and a sincere adherent of their promotion), supports this position and says "human rights standards will provide substantial guidelines in emergency situations only if they are specific about which rights can be infringed in emergencies."[1] Nickel finds The European Convention on human rights to be most appropriate and applicable, vis-à-vis the Universal Declaration of human rights (1948) that is silent on this issue and International Covenant on Civil and Political Rights, which leave most of human rights open to infringement when a necessary test is satisfied.

Nickel thinks that human rights can be violated in order "to protect national survival and security, during a period of extreme stress."[2] He acknowledges that such situations are problematic for human rights theories, since "they present both reasons for infringing human rights and conditions in which it is easy to exaggerate dangers and make mistaken judgments."[3] However, he still thinks that there are grounds for infringing upon basic rights such as the freedom of movement, the freedom of choice of residence, the rights to property and freedom for forced labor. He even feels that the right to due process can be infringed during national emergencies.[4]

An approach that acknowledges situations of national emergency as possible justifying grounds for the violation of basic rights could support, at least to some extent, the first case in this chapter. However, the acceptance of such an approach nevertheless cannot provide a carte blanche for violating basic rights, including the right to privacy, definitely not a comprehensive violation as allowed above.

CASE 1

The first example I want to deal with is a request, raised at the beginning of the current millennium, by the Intelligence Departments of the British Police and the two British security and intelligence services, MI5 and MI6, respectively. These agencies asked the British Ministry of the Interior to initiate legislation that would compel companies that supply electronic communication services to document and keep records of e-mails and Internet transmissions for a period of 7 years. The official reason for this demand was that sophisticated criminals, pedophiles, terrorists, white-collar criminals, and drug dealers use the Internet for promoting their malicious intentions. However, since they cannot determine in advance who is involved in such illegal activities, they need the record of all users.

Such a demand was a drastic change from the usual norms where warrants for using electronic apparatus for tracing information (such as telephone tapping) were issued in Great Britain under very limited circumstances, and only with regard to offenses that were already committed or in the planning stages:

---

[1] Nickel James. *Making Sense of Human Rights*. University of California Press, Berkeley, 1987, p. 131.
[2] Ibid.
[3] Ibid.
[4] Ibid, pp. 142–143.

(a) The offense must be really serious;
(b) Normal methods of investigation must have been tried and failed or must, from the nature of things, be unlikely to succeed if tried;
(c) There must be good reason to think that an interception would result in a conviction.[5]

This new request introduced a new dimension to the use of technology for invading people's privacy, since it applied to every user of the Internet and not only to confirmed suspects. Even before the Internet Era, people had fears about the ability of the authorities to impinge on people's freedom due to the immense amount of information they retain. Collin Mellors, for example, quotes Michael Harrington who refers to "The Bureaucracy" as "an enormous potential source of arbitrary impersonal power which folds, bends, spindles and mutilates individuals, but keeps IBM cards immaculate."[6] And this was in the 1970s, before computers became so powerful they could retain almost unlimited amounts of information.

The immediate response of human rights activists and organizations was an outcry that such a rule grossly violates the secrecy of personal information and the elementary rights of the individual to privacy. Their main wrath was directed against the attacks on one of the most fundamental aspects of privacy: the individual "possessing control over access to herself,"[7] including control over access to personal information about the individual. They demanded the outright rejection of this initiative and planned to struggle against this law with all the legitimate means at their disposal including an appeal to the European Court of Human Rights, should the British Parliament confirm such legislation. They considered such a comprehensive monitoring of people's correspondence as one of the most intrusive forms of violating the right to privacy. Charles Fried who examined the question of privacy in the light of the availability of monitoring devices (again, in the 1970s, where the eavesdropping apparatus and opportunities were much less effective than today) arrived at some ominous conclusions about such monitoring. He says that a system of monitoring conversations and other data (and this is the kind requested in the above example) "drastically curtails or eliminates altogether the power to control information about oneself."[8] Fried assumes that privacy is important as a medium for most kinds of relations that require a degree of intimacy, and monitoring makes this impossible especially in such intimate relations as love and friendship. Furthermore, according to Fried, monitoring

---

[5] Todd R. W. "Electronics and the Invasion of Privacy." In: Young John B. (ed.), *Privacy*. John Wiley & Sons, Chichester, 1978, p. 310. There Todd cites from *Interception of Communication* (Birkett Committee), Cmnd. 283, HMSO, 1957.
[6] Mellors Collin. "Governments and the Individual—Their Secrecy and His Privacy." In: Young John B. (ed.), *Privacy*. John Wiley & Sons, Chichester, 1978, p. 98.
[7] Inness Julie C. *Privacy, Intimacy, and Isolation*. Oxford University Press. Oxford. 1992. p. 62.
[8] Fried Charles. "Privacy: A Rational Context." In: Wasserstrom Richard A. *Today's Moral Problems*, Second Edition. Macmillan Publishing Co, New York, 1979, p. 373.

undermines the capacity to enter into relations of trust.[9] Having these conclusions in mind, the human rights organizations and activists were appalled at the suggested legislation.

The demand for initiating this legislation did not appear in vacuum. The availability of apparatus for tracking and eavesdropping on one's communication on the one hand, and the relatively low risk involved in the operation of these instruments on the other, create an enormous temptation for those who want information of any sort. Here we refer not only to governmental and official institutions—which justify their demands on the grounds of national security, public order, or public peace—but also to economic and business agents, or even private investigators that work for private people. They all use advanced technological instruments for tracing phone calls or fax massages, photographing or recording other people without their consent or without an authorization from a legal-competent authority. By doing so, they grossly invade the privacy, autonomy, and personal space of the individual. Such proposed legislation strikes at one's privacy, reducing the extent to which an "individual is afforded the social and legal space to develop the emotional, cognitive, spiritual, and moral powers of an autonomous agent."[10]

We do have the ability and legitimacy to disallow private agents (such as private investigators or insurance agents) from invading people's privacy and we can prosecute them or at least publicly condemn them and their actions if they continue. However, an invasion that is conducted by the government, which represents "the single biggest collector and distributor of information about citizens,"[11] can establish legitimacy for such intrusive actions. Such legitimacy may lessen the moral obloquy of such an invasion when it is done illegally by private agents, since it does not appear essentially different from governmental invasion into people's privacy. Thus, one's immunity against external invasion of one's privacy—at least with regard to one's private life and personal sphere—becomes worthless if not meaningless. This is because private offenders can claim that since the invasion into one's privacy by the police or by governmental agents does not undermine one's dignity, humanity, and autonomy, then their own actions are not that serious either.

The main difficulty in protecting one's privacy, particularly against governmental intrusion, is convincing both people and governments that just as we are not required to divulge our health record, bank account, or room number in a hotel to a stranger, it is also not legitimate to require us to do so for governmental authorities. When this issue becomes clearer, it will be easier to recognize the absurdity of the demand of the intelligence authorities, and it will be possible

---

[9] Ibid.
[10] See, Schoeman Ferdinand David. *Privacy and Social Freedom*. Cambridge University Press, Cambridge, 1992, p. 13.
[11] See Hixson Richard F. *Privacy in a Public Society*. Oxford University Press, Oxford, 1987. p. 207.

to increase the penalties of those eavesdroppers or snoopers. In order to do so, we can look at the analogous situation of protecting rights against governmental cupidity and unlimited appetite, which would clarify the immunity that should be given to the right to privacy.

I hold that one's right to privacy has as much weight as one's right to property. The right to possess property has become almost sacred in modern society, and the protection against its violation is virtually the only indisputable right in capitalist societies. The common ground for these two rights may be found in the fact that both rights assist in the foundation and establishment of one's subjectivity and autonomy: firstly as a possessor of will, desires, aspirations, and unique preferences and secondly as a possessor of property. This individual subjectivity requires that a human being be seen as a free and autonomous agent.

The idea of connecting privacy rights to property rights was raised by Judith Jarvis Thomson, though in a slightly different manner from the one I want to suggest here. She does this as a simplifying hypothesis and argues that the right to privacy is itself a cluster of rights, which is not "a distinct cluster of rights but itself intersects with the cluster of rights which the right over the person consists in and also with the cluster of rights which owning property consists in."[12] I follow her idea only in the sense that the right to privacy should have the same inviolability that modern society ascribes to the right to property.

My idea here is not to argue that the right to property carries "an implied right to privacy"[13] or to say that we should value privacy because it provides us with liberty or control over our property. Here I fully agree with Julie Inness, that privacy claims are "conceptually distinct from liberty claims or property claims."[14] I only suggest here that the same immunity, respect, and weight we ascribe to property rights, should be given also to the right to privacy, as something that protects not only our intimacy but also our subjectivity and autonomy (and this is common for both privacy rights and property rights). What I want to stress is that this subjectivity, as the core of a person's autonomy, is also the core of any personal claim, including the claim for the possession of property rights.

Authorities have already come to acknowledge individual property rights as inviolate because of their respect for the individual as a rational and free person whose self interests might be different from the interests of the collective or interests of the authorities. If they would also internalize the idea that a person's freedom and autonomy include not only property rights, but the right to privacy as well, they might not raise such all-inclusive demands for striking at the people's right to privacy. When the authorities begin to respect the imperviousness of

---

[12] Thomson Judith Jarvis. "The Right to Privacy." In: Schoeman Ferdinand D. (ed.), *Philosophical Dimensions of Privacy*. Cambridge University Press. Cambridge. 1984. p. 281.

[13] Such an idea was raised by two Boston Lawyers in the case of *Roberson v. Rochester Folding Box Co.* in 1902. See Hixson Richard F. Ibid, p. 39.

[14] Inness Julie C. Ibid, p. 139.

persons from the transfer of personal information to others, or even against the retention of such information, we will also be able to fight against the transfer and use of information made by private agents as well. If the bill that was proposed by the British intelligence authorities would be legislated, it would likely realize George Orwell's dreadful prophecy of a world where the citizens' lives would be persistently controlled by the authorities and they would have no more real freedom than prisoners in a penal institution. In the face of this danger, the fear that criminal elements may use the Internet for their own aims (the original reason for the initiative) is minor compared to the fear that the authorities will access and use such personal and intimate information. This is because criminal elements have a much more limited ability to illegally exploit the Internet while government authorities have omnipotent abilities to misuse the Internet without any way of deterring them. My conclusion is that we should fight the criminal abuse of the Internet only in a way that will not harm or invade the privacy of innocent users.

Such an attitude towards the concept of the right to privacy follows Ronald Dworkin interpretation of this right. Dworkin says that this concept "argues that people should have a private sphere even if this damages rather than advances society's long term goals."[15] One's privacy, when it may be threatened with an all-out attack, should be protected as if it were a threat on one's personhood and autonomy. It should override the state's interests, even when security considerations are involved.

Dworkin's ascription of the priority to one's privacy over social goals expresses a position that considers the right to privacy as more fundamental for the establishment of one's personality than any other right, including the so called "sacred" property rights. Jeffrey H. Reiman, who criticizes Judith Jarvis Thomson's idea of comparing the right to privacy to property rights, believes that "the right to privacy is the right to the existence of social practice which makes it possible for me to think of myself as the kind of entity for whom it would be meaningful and important to claim property and personal rights ... Thus the right to privacy protects something that is presupposed by personal and property rights."[16] Reiman thinks that privacy has a significant role in the establishment of one's individuality, which is itself the ground for one's claim for property rights. He argues that "Personal and property rights presuppose an individual with a title to his existence—and privacy is the social ritual by which that title is conferred. The right to privacy, then, protects the individual's interests in becoming, being, and remaining a person."[17]

---

[15] Dworkin Ronald. *A Matter of Principle*. Harvard University Press, Cambridge, Massachusetts, 1985, p. 350.
[16] Reiman Jeffrey H. "Privacy, Intimacy and Personhood." In: Wasserstrom Richard A. (ed.), *Today's Moral Problems*, Second Edition. Macmillan Publishing Co, New York, 1979, p. 391.
[17] Ibid.

Reiman's view which is close to the one I want to promote here, shows that the right to privacy lies at the very core of a person's self identity and individuality. And if we consider that privacy rights are fundamental for being able to make claims for property rights, we would think of the right to privacy as having at least the same weight we ascribe to the right to property, if not greater. My argument here is that since the right to privacy is no less important than the right to property, and the latter has long been accepted as immune and invulnerable from needs of the state or even its rapaciousness, we should start promoting the right to privacy as deserving at least the same respect and immunity. Such an approach rejects the idea of recording electronic transmissions about all citizens, just as it would reject the demand to confiscate everyone's bank accounts simply because some criminals and terrorists use their bank accounts to finance their activities. This parallelism demonstrates the absurdity of such a comprehensive demand by the intelligence services.

At this point I want to go back to the idea raised at the beginning of this chapter and ask a slightly different question: could the authorities use the approach supported by James Nickel about the legitimacy of violating human rights, as a basis for demanding that Internet companies keep records of the users' correspondence? Nickel argues that such violations of privacy are allowed only in situations of national emergencies, and as examples of these he lists wars, disasters, armed foreign invasions, earthquakes, and insurrections. Since even Nickel restricts permission to violate human rights to very specific circumstances of extreme stress, it is clear that these do not include violating private correspondence during daily life. So the answer to our question here is strictly negative. However, the question remains as to whether this approach can justify the original request to trace and eavesdrop on people's electronic correspondence during national emergencies. This is a more complicated question since one can claim, as Nickel does, that during national emergencies "governments have the job of minimizing the damage to people and property, restoring order and security, and repairing the most disruptive damage. In order to do these things, certain emergency powers are often necessary."[18] And in order to protect national survival and security, we should be able to use harsh measures.

Here I am less unequivocal than before, but believe that even under such circumstances we still have to tread carefully. I agree with Harvey Mansfield that during national emergencies: "it is especially difficult to protect everyone's rights. But then, one might reply, it is obviously just when one's rights are in danger that protection is most needed. It seems, therefore, that rights are most precarious in times of emergency when they are most needed."[19] Complying with the demand of the Intelligence services, even if only during national emergencies, may lead

---

[18] Nickel James. ibid, p. 142.
[19] Mansfield Harvey C. "Human Rights in Emergencies." In: *Critical Review*, Vol. 6, No. 4, Fall 1992, p. 575.

to the annulment of the right to privacy during daily life. If we allow the violation of rights when they are most needed, we reduce their value when they are less needed. Therefore, the original demand—the monitoring, tracing, and recording of every piece of information about every user of the Internet—should be rejected even during national emergencies, since it is indiscriminate. If it was limited to records of suspects involved in crime or terrorism, the intelligence services could get permission from an authorized institution or a judge and would not need any further legislation. When the demand is to record every electronic transmission, we cannot comply with such a demand even during national emergency.

To sum up this example I want to stress that the interests represented by intelligence services cannot justify the blanket request to intercept every electronic message that passes through the Internet. Almost all the users whose messages could be traced are not suspected of any criminal or terrorist activity, and there is no reason to track them. Certainly if there are specific reasons to suspect someone, a judge can authorize the tracking of his/her massages. Treating the entire population as if they are enemies of the state, brutally strikes at the fundamental grounds of their freedom, autonomy, and even their subjectivity and personhood. Thus, there is no place for such a comprehensive demand at any time, not even during national emergencies. Even those who think that the right to privacy does not protect autonomy directly, such as Carl Wellman, nonetheless believe that there is a solid connection between autonomy and privacy, albeit indirectly. Wellman thinks that though the right to privacy does not directly protect autonomy, it still protects "constitutionally recognized areas of privacy from invasion and thereby protects personal autonomy indirectly by providing areas within which individual decisions and action will be free from intrusions that would damage or destroy autonomy. Moreover, these zones of privacy may well be constitutionally recognized for that very reason."[20] Wellman's position would lead to the same conclusion as mine: that such an intrusive blows at one's privacy harms one's autonomy, and cannot be allowed. Wellman even thinks "the personal and social value of autonomy probably figures predominantly in the grounds of the right to privacy."[21] Thus the government cannot raise such demands without undermining people's autonomy. Hence, such demands should be rejected, even according to Wellman's position.

CASE 2

The second example is similar to the previous one, but has aspects that may require a slightly different attitude towards it. This concerns the persistent demand to establish a DNA information bank or database that may help in spotting

---

[20] Wellman Carl. *An Approach to Rights*. Kluwer Academic Publishers. p. 186.
[21] Ibid.

criminals whose biological data and details were found at the crime locations. Here the demand is slightly different from the idea of establishing a DNA database for transplant or other medical uses. The inclusion of people in a medical-type database is voluntary and mutually beneficial, since they may benefit from such inclusion if they themselves ever need a transplant, while a DNA bank would be a compulsory database that includes anyone who has ever been arrested or even questioned by the police. Those who support the establishment of such a database argue that such an information bank would facilitate the tracking down of many undetected criminals (particularly sex criminals and offenders) since many of them are recidivists and their DNA details would be included in that database after their first arrest. Those opposing this idea are concerned about the additional blow to citizens' privacy, especially to those arrested and later found innocent. They consider this demand as a further invasion of the authorities into a new sphere, which should be protected from external control and inspection.

The significance of the right to privacy has already been discussed in the first chapter of this book and the first part of this chapter. One's right to be let alone has been widely acknowledged and justified, so I presume its importance as obvious. Those who object to the idea of a DNA database for fighting criminals are concerned about the possibility that domains currently protected by the right to privacy, such as health records and genetic makeup (including genetic deficiencies) would be revealed to the authorities and retained in databases. They understand the right to privacy as respecting a person's individuality and subjectivity. According to the defenders of this right, the authorities are not allowed to eavesdrop on people's phone calls, or to search people's homes and offices without having proper authorization and a special search warrant from a judge. Even then, the judge has to first make sure that these actions are essential for preventing a crime or for assisting in the investigation of a crime that has already been committed. The authorization and search warrant are limited to a specific use, and not for any other legal purpose. Even then, those who are concerned about people's privacy and autonomy are not very happy about the ease and simplicity of obtaining search warrants in general.

When we examine the demand for establishing such a database in light of the right to privacy, the most relevant factor is the measure of the possible harm to the individual that may result from their being included in that database. If we can minimize this harm and still benefit from the existence of that database, it would justify its establishment. Of course, one has to be very strict with preserving the safeguards and restrictions imposed on its access and use. We should make efforts to minimize unnecessary harm that may result from such a database, such as stigmatizing those included in it. Such a concern may be significantly reduced if we include as many people as we can in that database, such as not only those suspected of criminal behavior but also members of other neutral groups. This may not only reduce the stigma of those included in it, but may increase its efficiency since we can never know who may be involved in criminal activity. A database that is similar to the one described here is that of fingerprints: to cite just two examples, fingerprints are taken from everyone who is recruited into the Israeli army and every foreigner entering the US. Being included in such a huge database is not incriminating at all, since there is no stigma to entering the US, for example. Another fear is that of the abuse of the genetic substance that will be taken from the tested persons. This fear may be reduced by ensuring that the samples will be destroyed immediately after their analysis, and allowing only the existence of a computerized database and not of genetic bank that actually stores the DNA samples. Thus, possible misuse of human genetic substance will be limited to a very short time span (i.e. the time it takes to map the sample and store the result in a computer).

However, a general computerized database still contains intimate personal information about individuals (usually protected by medical secrecy). There is still the fear that exposing such information will cause harm to those included in the database, even if this harm involves only an attack on their right to privacy. We have to make efforts to minimize this fear as well. We can do it the same way we protect privacy when we allow the authorities to invade someone's home only when a judge authorizes a search warrant that restricts the search to the specific aims requested by the authorities. Access to the database should be restricted and should be on the same level as access to one's home or one's phone line: both should require authorization from a judge.

I should be stress here that in principle, maintaining such a database does not constitute a serious moral problem. Such a database may add some new information to the immense amount of information that the authorities already possess (age, address, living standards, income, education, etc.). An acute moral problem may occur as a result of misuse of such a database, thus severe restrictions should be imposed on its misuse and harsh penalties meted for its violation. Perhaps under different circumstances, if there were fewer crimes and greater resources allocated to fight them, we could relinquish the idea of such a database and obviate the fears of abuse from such information. However, if the absence of such a database allows criminals to continue harming innocent victims for years to come, we should consider the establishment of a DNA database for fighting crime, especially for reducing recidivism. Of course, its use should be restricted to fighting crime and terrorism. Perhaps relinquishing the idea of such a database is a luxury which modern society cannot afford. In this connection, Collin Mellors was correct to assume that "Criminal and medical records are two other areas where it is generally wise to retain data. The retention of data is not, in itself, necessarily harmful. The key questions are who decides which information is to be retained and which discarded, and how any which is retained should be used."[22]

---

[22] Mellors Collin. Ibid.

Public security vs. the right to "be let alone" 35

We still cannot remain unperturbed against the threats imposed by information banks, even if they are only computerized. D. W. Barron divided the threats to the individual from a mechanized databank into three main forms:

(i) Illicit access to information with malicious intent by persons not entitled to such access.
(ii) Unexpected consequences of making information freely available to authorized persons by mechanical means.
(iii) The use of information for purposes other than that for which it was collected in the first place.[23]

We will restrict ourselves here to *direct* threats to privacy, though there are indirect ones as well. Relevant to our discussion are Barron's suggestions of a "Data Protection Committee" which were considered by the "Young Committee." These suggestions can serve as safeguards to reduce privacy risks when establishing a database or can serve as guidelines for the more general problem of information secrecy. Those recommendations were as follows:

1. Information should be regarded as held for specific purposes and not be used, without appropriate authorization, for other purposes.
2. Access to information should be confined to those authorized to have it for the purpose for which it was supplied.
3. The amount of information collected and held should be the minimum necessary for the achievement of a specific purpose.
4. In computerized system that handles information for statistical purposes, adequate provision should be made in their design and program for separating identities from the rest of the data.
5. There should be arrangements whereby the subjects will be told about the information held concerning him.
6. The level of security to be achieved by a system should be specified in advance by the user and should include precautions against the deliberate abuse or misuse of the information.
7. A monitoring system should be provided to facilitate the detection of any violation of the security system.
8. In the design of information systems, [time] periods should be specified beyond which the information should not be retained.
9. Date held should be accurate. There should be machinery for the correction of inaccuracy and for updating of information.
10. Care should be taken in coding value judgments."[24]

---

[23] Barron D. W. "People, Not Computers." In: Young John B. (ed.), *Privacy*. John Wiley & Sons, Chichester, 1978, p. 320.
[24] Ibid, p. 327.

Not all these principles are relevant to the DNA database discussed here, but they give us general ideas about the restrictions we should impose on such as database. We must monitor those who have access to any information held in a DNA database. Barron reminds us "one powerful weapon will be an increased public awareness of the potential dangers and benefits of computerized databanks."[25] Our responsibility as citizens is such that if a database is created, we must stay on alert to make sure that it is monitored appropriately and not abused; for example, that it is not made accessible to insurance companies or private individuals with vested interests.

One may ask why I am so decisively against collecting electronic Internet information about people on the one hand, while on the other hand I reluctantly accept the need for establishing a medical database. I think that there is an important difference between the two situations. The information in a DNA database is very specific and limited to biological data, and does not include any other personal information. The electronic information transmitted on the Internet, on the other hand, may include such information as social, economic, personal, and mental data. We cannot allow such comprehensive monitoring of citizens through the use of such information and still consider ourselves to be free agents. Such comprehensive monitoring would totally eliminate our right to privacy, while the collection of only biological information (whose access would be restricted) does not represent such a brutal intrusion into one's social life. This is only one aspect of people's privacy, and though very intimate, it has fewer implications for social and public interactions. The benefits we may gain from such a medical database may partly compensate for the unavoidable blow at our privacy.

To sum up this chapter, I want to say that we should be extremely skeptical about the establishment of more and more databases that collect and retain information about people. There are cases where such databases may be justified but even then, we must be aware of the dangers involved and take proper precautions. These precautions will allay the public's fears of a "Big Brother" Orwellian state in this computerized era, and show deep concern for an important human right: privacy.

[25] Ibid.

Chapter 4

FREEDOM OF EXPRESSION IN ACADEMIA AND THE MEDIA

Freedom of expression has become "a highly regarded concept around the world."[1] Its value has become so common and acknowledged in the western liberal countries, that Robert Trager and Donna Dickerson even argue that "Americans have a tendency to believe that the freedom to express their beliefs is a basic part of what it means to be an American."[2] Even though its breadth and depth differ from country to country, more than 60 countries have freedom-of-expression provisions in their constitutions.[3] The fact that freedom of expression is included in so many constitutions and in every universal or international document on human rights, "says that there is some level of consensus about its value."[4] However, its content, boundaries, and practice are not very clear, and sometimes even vague and disputable. There is always tension between our desire to maximize freedom of expression and between other considerations, some of which are moral or normative. The tension between social and political desires to intensify freedom of expression and aesthetical and ethical considerations become very complicated vis-à-vis the media and academia. Both domains usually enjoy some extent of immunity against censorship, and we are very careful to avoid limiting the range of their freedoms and particularly the freedom to broadcast things that we basically loathe. We believe that the independence of editors, reporters, and university professors is beneficial for society, and their autonomy is a significant part of their

---

[1] Trager Robert, and Dickerson Donna L. *Freedom of Expression in the 21st Century*. Pine Forge Press, Inc. Thousand Oaks. 1999. p. 92.
[2] Ibid, p. 91.
[3] Ibid, p. 92.
[4] Ibid, p. 98.

journalistic mission and their creative activity. In college campuses, academic freedom has became almost sacred and we will not tolerate the possibility that political or economical interests or considerations would affect the curricula or the contents of courses given in academic institutions. We think that academic freedom is crucially important for the progress and creativity of scientific work, and hence ascribe full autonomy to research teams in the universities as well as to the lecturers.

This chapter examines issues of freedom of expression in the academia and the media, in specific contexts that are purposely abhorrent to most of us. It is when this freedom is used to air repugnant opinions or abhorrent televised news footage, that these can be viewed as "test cases" to examine the limits and boundaries of our acceptance of freedom of speech.

We may discover that even though we strongly reject external censorship on the media, we sometimes expect them to operate self-censorship and abstain from publicizing certain views or pictures or films which we consider to be inappropriate. We believe that it is the media's duty to report and cover every issue, which the editors regard as having public interest, but the way they do so should be intelligent and relevant. In the first part of this chapter I deal with two examples, which appear to me as inappropriate attempts to use the media for malicious aims, and argue that these are the sorts of cases where the media should impose restrictions and limitations on themselves.

The case of academic freedom is a slightly different issue. We respect it as something that possesses intrinsic value, and we think that it is a unique privilege that those who work in academic institutions should enjoy, due to their special status as researchers. Even though we know that this freedom can sometimes be abused by lecturers and researchers, we are usually tolerant toward this abuse, since we are more afraid of its curtailment by external authorities or interests. The second part of this chapter deals with an example where we might think that we should restrict this freedom. However, even in the unusual case with which I deal in that part of the chapter, I will argue that we should respect freedom of expression, even if, as in this particular example it irritates us and we think that this specific use of academic freedom is outrageous.

CASE 1

In the first part of this chapter I examine the question of the airing of appalling footage in the public media of television. I deal here with two cases, which occurred more or less at the same time—during the year 2000.

The first case was the broadcasting of a videotape on television that was photographed by a rapist, in which he documented the cruel abuse and torture of his victim during the attack and the rape. The videotape was shown in the main news edition of the Israeli Broadcasting Authority—Channel 1, which is shown daily at 9:00 pm and attracts a wide audience consisting of diverse sections of the public. Before broadcasting the tape the anchorperson warned that the next item would be something appalling and hard to watch. The tape showed a naked woman who was attacked and cried out, begging unsuccessfully to her assailant to stop hitting her and leave her alone. The broadcasting authorities justified the screening of such a tape on prime-time news by declaring their desire to draw the public's attention to a horrible phenomenon that might well become more frequent in Israeli society.

The second case concerned the request made by Timothy McVeigh who was convicted and condemned to death for committing the April 19, 1995 terror attack in Oklahoma city, in which 168 people were killed. He asked that his execution should be televised and shown on public TV so that everyone would be able to view it. He argued that since the authorities intended to show the execution on closed-circuit television anyway, to a crowd of about 250 people who had survived the terror attack, the principles of justice and equality required the screening of the execution in front of the entire population, or at least in front of everyone who was interested in watching.

I do not want to delve too deeply into the reasons behind the original decision to screen the execution in front of the survivors. However, I confess that I cannot fathom any benefit to be gained by viewers, even survivors of a terrible terror attack, who watch the execution of so abominable a murderer. At first glance it appears like an attempt to satisfy their desire for revenge. But if this is indeed the preference of the survivors, and the authorities wish to respect this desire, the assassin's right to privacy is not a real consideration which should lead the authorities to reject their will or request. Viewing the execution might have some curative or remedial value for the survivors, and they might gain some benefit, even though I cannot see now what this might be. In any event, the assassin's request may be worthy of further examination.

It is clear that the rape videotape and the screening of the execution are dramatically different from each other, both in their essence and in the purposes for which they are broadcasted. However, they both raise the same questions: Is everything worthy of screening and broadcasting? Aren't there any limitations or restrictions that should be imposed on public viewing of certain things? It is obvious that the media which want to broadcast such events can say that the public is very interested in these cases, and hence it is their duty to fulfill the public desire to watch these tape. They can maintain that it is not their duty to censor the pictures for the public, to educate the public or refine its tastes and that the public can decide for itself whether to watch it or not. A less obvious query is whether there is any news or journalistic added value to the broadcasting of these events. From the point of view of the public's right to know, it is enough to report the events verbally and avoid televising them visually. Thus the claim of public interest does not justify the public broadcasting of such horrendous pictures.

However, the representatives of the electronic media can claim that they are not merely press services, which are supposed simply to provide the public with information, but also (or even primarily) entertainment organizations whose mission is to pander to the tastes and demands of the public. They will probably, nonetheless, not be so ready to admit that they are *primarily* business ventures which aim at producing profits for their owners. Thus, if there is a demand or desire for watching such terrible scenes, the electronic media as an entertainment vehicle will attempt to satisfy this demand for pecuniary reasons. This may explain, but not legitimize, the frenzy for ratings and popularity, which sometimes appears to be the only consideration that determines what will be broadcasted and what will be passed over in silence. From this point of view, which places the rating criterion at the top of its considerations, the broadcasting of events such as those mentioned above seems to promise a huge success. The adherents of this approach toward the electronic media may claim that the responsibility for the contents of the TV programs relies basically on the public taste, and the media merely reflects this taste.

But the consideration of the electronic media as merely entertainment instruments, which are supposed to suit the content of the broadcasted programs to the public taste, and the attempt to remove any moral or ethical responsibility for the contents of its programs, appears to me to be a fraud and a shirking of their responsibility to the public.

The main reason for my rejection of this approach is the obvious fact that the TV and radio channels do not merely follow the public's taste and its preferences, but to a large extent *form* and *determine* the public's preferences and taste. The electronic media (as well as the press) has an educational responsibility to the public of which it must be fully conscious. The broadcasting of horrific events and spectacles increases the measure of violence to which the general public is exposed, reduces and dulls the public's sensitivity toward such phenomena, and blunts the loathing and condemnation that, morally, we ought to feel. People's preferences are significantly created and formed by what they watch, experience, and become used to. The electronic media are very influential in forming the public's taste as they have become the most accessible vehicles of public communication. Frequently, people's preferences are merely the adoption of what they see on TV. The media cannot therefore hide behind the claim that "this is what the public wants." The media themselves bear full responsibility for the regrettable lapse in the tastes and proclivities of the public. The media can and ought to assist in stopping the deterioration of the public taste. Instead of lowering the level to a deeper abyss by broadcasting and showing these horrible spectacles, they should pointedly abstain from doing so for the good of the public.

The other claim, behind which the leaders of the Israeli national channel (Channel 1) tried to hide, was their desire to shake the public consciousness in order to galvanize public action that might uproot such phenomena from Israeli society. This claim sounds ridiculous, since those people who feel solidarity with the victim are the same people who already loathe such crimes and object to giving publicity and a public stage to such offenders or to their acts. These people are sufficiently shocked and aghast that they do not need additional terrifying spectacles and visuals in order to understand the gravity of such crimes. They do not need an external trigger to move them to action, since their attitude toward such crimes and criminals is solid and determined. Other people, with vulgar tastes and sadistic tendencies who want to see more of such scenes, will not be shocked or shaken; on the contrary, these images which only increase their sadistic and violent impulses and tendencies. They will definitely not change their attitude toward such crimes, and there is even a threat that the opposite may happen. Some viewers may be encouraged to imitate what they see on TV. When the Israeli Broadcasting Authority prefers to pander to the tastes and demands of these people, it taints and tarnishes its own aesthetic preferences. Here I can only accept the position of Raphael Cohen-Almagor who thinks that "The media have certain obligations to fulfill. They should be fair and not exaggerate, view people as ends rather than means to something, take into account the consequences of reporting, reveal what is reported, and not refrain from making distinctions."[5] The media should heed Cohen-Almagor's words when they decide what to report verbally and what to televise visually.

The Israeli Broadcasting Authority functioned in this instance as if it were a local and private TV station in the US, even though this is a government channel (which should not be affected by commercial considerations), and a national one (which should be more sophisticated and moderate). According to Shanto Iyengar, "Local news is defined by a distinctive perspective on public issues and events, that is, by its emphasis on (and frequently exaggeration of) drama, conflict, and violence. Every effort is made to appeal to the public's appetite for "blood and guts." All local broadcasters are well aware that if local news is to be economically successful, it must emphasize violent crimes."[6] However, this may cause the banality of evil, as Mark Kingwell, following Hannah Arendt, describes such a process. He argues that "Evil becomes banal ... when it becomes mechanical, routinized, heedless, and thick."[7] When violence is shown daily on TV news it becomes routine, common, and banal. Mercy toward the victims is replaced by our inquisitiveness and curiosity, and we might end up in a situation where we search for such news and events to satisfy our perverted tastes. The most harmful consequences of television result from the screening of gory and

---

[5] Cohen-Almagor Raphael. Speech, Media and Ethics: The Limits of Free Expression. Palgrave. 2001. p. 72.

[6] Iyengar Shanto. Media Effects: "Paradigms for the Analysis of Local Television." in: Chambers Simon, and Costain Anne. (eds.). *Deliberation, Democracy and the Media*. Rowman and Littlefield Publishers, Inc. Lanham. 2000. p. 108.

[7] Kingwell Mark. "The Banality of Evil, The Evil of Banality" in: Chambers Simon, and Costain Anne. (eds.). *Deliberation, Democracy and the Media*. Rowman and Littlefield Publishers, Inc. Lanham. 2000. p. 185.

violent footage on national prime-time news. Colin Munro says that "the visual medium of television, being more powerful than the written word, may have an undesired impact or unmerited influence on attitudes or behavior."[8]

Here we can clearly see the conflict between "good journalism" and "good stories," where often one "good" comes on the expense of the other. Raphael Cohen-Almagor believes that when this happens the need to produce a "good story" often prevails. He asserts that "Good journalism, which involves the requirement of objective reporting, might become no more than a token, sometimes to which journalists pay lip service. After all, good stories (which are often deal with phenomena such as terror, war, drug addiction, rape, violence, and racism) are more likely to sell newspapers and boost ratings."[9] Cohen-Almagor explains the difference between reporting events verbally and televising them visually in the Israeli media when he examines a controversial event that occurred on October 1995 in Zion Square in Jerusalem. There was a demonstration there that protested the Oslo Accord as well as the Rabin Government that signed this accord, and some people at that demonstration carried posters in which the Israeli Prime Minister, Yitzhak Rabin, wore the uniform of the Nazi leader Heinrich Himmler. The leading Israeli newspaper, *Yedioth Ahronoth*, reported the existence of such posters but did not show or print them, saying that "such pictures were too objectionable and did not deserve publication."[10] This is unusual since the competing newspaper *Ma'ariv* did publish those pictures. Cohen-Almagor justifies the decision not to print the picture by saying that the newspaper did its duty in reporting the event but at the same time "did not serve as a promoter of hatred and incitement."[11] This distinction is extremely significant for editorial considerations in general, and most relevant for the case we discuss here. Televising tapes, which were filmed for nefarious purposes, may promote the photographer's evil intentions and aims. In the case here, the broadcasting authorities functioned (unintentionally) as the promoter of these appalling aims and intentions.

The analysis of the case of the Israeli Prime Minister's portrait wearing the Nazi uniform enables us to reflect on and reassess the other claim used by the Israeli Broadcasting Authority, that the screening of the rape may help in drawing attention to the terrible phenomenon of rape in the country. The same justification could have been used by the newspaper, which printed that photomontage. The newspaper could argue that the printing of that picture was "increasing public awareness regarding the phenomenon of hatred, and arguably creating a much more intense public reaction to the level of hatred against a designated individual or government".[12] Truly, it is much stronger to show the picture than to merely

---

[8] Munro, Colin R. *Television Censorship and the Law*. Saxon House. Westmead. 1979. p. 171.
[9] Cohen-Almagor Raphael. ibid, p. 78.
[10] Ibid, p. 91.
[11] Ibid.
[12] Ibid.

report on its existence. However, it can also boomerang and intensify the feelings of hatred among those who oppose the government. Cohen-Almagor concludes his discussion of that event, "Printing such a photomontage fuels an atmosphere of incitement against the designated target. This atmosphere was conducive to the event that took place on 4 November 1995: Prime Minister Rabin's assassination."[13] I do not want to draw a comparison between the two events, but only to claim that eschewing "good journalism" for the sake of a "good story" only causes damage to the publication itself.

Another issue that relates more to the second case, that of Timothy McVeigh, but is also relevant to the case of the rape, is worthy of being mentioned here. If those who should decide whether or not to broadcast such tapes cannot understand the deep moral and normative implications of publicly screening such tapes, they should ask themselves who has interests in broadcasting them, and what these interests could reflect. The tape that was publicized in the Israeli news edition was filmed by a violent rapist, possibly for the fulfillment of his own sick and sadistic impulses, or to boast about what he had done to other iniquitous people like him, or in front of people who have the same perverted tendencies. If his original aim was the satisfaction of his evil impulses, the broadcasting of this tape may benefit people like him: exactly the same people against whom any medium should fight. If his aim was to glorify himself in the eyes of other people, that is all the more reason why the media should not give him the satisfaction of providing such a large audience to his perversions. Any compliance with this desire is mere complicity with evil desires or perversions.

The understanding of this point is easier in the case of the murderer, since the broadcasting of his execution is not only compatible with his desire and will but also is a fulfillment of his explicit request. This might be considered as a reward for what he had done, and thus is very good reason to reject his request. Even if the electronic medium believes that it has to publicize or report these events it should do so in a manner that harms the interests of heinous criminals and not in a way that promotes their interests, else the perpetrators come off as victims or, even worse, as heroes. The broadcasting of such dreadful scenes not only harms the feelings of the people who loathe these scenes, but also encourages the people who support such cruelty. In this sense the medium may even be accused of complicity with the brutality and violence against which it pretends to warn when it televises such brutal footage. And here I do not enter into the moral or normative questions concerning public executions (or capital punishment), but simply to the question of complying with a cruel request made by a cruel assassin. Even if we could show some utility to result from such a screening (for example, if we could demonstrate that it might have some deterrent effect on future criminals), we should reject such a screening on the grounds that such assassins do not deserve any glorification, even through their death.

---

[13] Ibid.

To conclude this section I want to stress my view that though not everything is worthy of being published or screened, I insist that there should not be any external censorship on television or newspapers. These media should be professional and impose self-censorship upon themselves. They should consider both the meaning and possible consequences and implications of any decision they take, and understand their role and responsibility in society. If they fail to do so, and take into account only ratings or commercial considerations, the public should keep them in check: not by legislation, which limits or restricts the freedom of speech and expression, but through civil action such as protest or even a consumer boycott. This could be more effective than legal censorship, and without the danger of reducing freedom of any sort through the law.

CASE 2

The second issue I want to discuss in this chapter is a debate which took place in the Hebrew University of Jerusalem about whether or not to allow one of its professors, Martin Van Creveld (an expert in military history), to teach a course on "Feminism Rejected." This course was given in the Department of History and in his lectures, Prof. Van Creveld expressed his views of the superiority of men over women and explained, according to his theory, why the feminist movement had failed and did not attain any significant achievements.

To some extent, the content of these lectures can be viewed as belonging to the category of "vilification" or "hate speech." Katharine Gelber, who investigated the issue of "hate speech," regards such a speech as "speech which is particularly harmful because it contributes to a climate of hatred and violence toward marginalized and disempowered sectors of the community. It violates the basic human dignity of its victims."[14] In this part of the chapter I examine the issue of that specific course as if it was one of the "hate speech" cases, and see if there is any justification for banning Prof. Van Creveld from the teaching of his course.

Apparently, the debate here is about academic freedom and its limits or restrictions. The usual approach to academic freedom requires that the institutions should not intervene in the contents and curricula of the courses, or the contents of the research studies of academic faculty who work in these institutions. The convention among academic institutions is that a faculty member possesses the prerogative of deciding what will be taught in his courses and classes, and what subject he/she examines as a researcher. The rationale behind this procedure is the assumption that academic freedom is essential for the progress of research, and indirectly, the progress of society as a whole (some may even say of civilization in

---

[14] Gelber Katherine. *Speaking Back: The Free Speech Versus Hate Speech Debate*. John Benjamin Publishing Company. Amsterdam. 2002. p. 1.

general). The freedom to teach whatever the lecturer wants is supposed to ensure that the students are exposed to a variety of different ideas, methodologies, and ideologies. From this multiplicity of ideas, students are supposed to select and form their own autonomous academic attitudes. However, many people (including academic figures) think that academic freedom should not be used as camouflage for the promotion of a lecturer's personal beliefs or positions. A lecturer should suggest "pure scientific" or "objective" truths, and not involve his personal positions, preferences, and tendencies. In the case discussed here, there was harsh criticism—not only within academic institutions but also among wide sectors of society—against Prof. Van Creveld's positions, and there were many people who blamed him for donning the academic robe in order to disguise the chauvinist and sexist opinions he wants to promote.

However, we can expand this debate to the more general question about the appropriateness of allowing the expression of ideas and opinions which are condemned by most parts of society (even though I am afraid that at least in Israel, such "conservative," "old fashioned" or should we better say "dark" ideas as those of Prof. Van Creveld are quite popular among large parts of society, and not only among the orthodox religious sectors). John Stuart Mill already dealt with this question in detail when he examined the question of "The Liberty of Thought and Discussion," in the second chapter of his famous book: *On Liberty*. Mill believes that we should tolerate unusual or unpopular opinions since none of us is immunized against making mistakes. Mill explains his position through four different arguments. First, there are cases where the dominant or politically correct opinion is mistaken or wrong, and the exclusion of other opinions and ideas might prevent us from recognizing the right or correct opinion. Mill says: "If any opinion is compelled to silence, that opinion, may for aught we can certainly know, be true. To deny this is to assume our infallibility."[15] Mill actually believes, "All silencing of discussion is an assumption of infallibility."[16] For me, this claim is enough grounds to reject all silencing arguments, since I reject any such assumption of infallibility. Another argument says that sometimes, particularly in the domains of politics and society, both the common opinion and its opposite are not sufficiently persuasive, and we might get closer to the truth only through a synthesis between those two competing opinions. The concealment of the opinion that contrasts the dominant or prevalent one might delay us from approaching the truth. According to Mill, "though the silenced opinion be an error, it may, and very commonly does, contain a portion of truth; and since the general of prevailing opinion on any subject is rarely or never the whole truth, it is only by the collision of adverse opinions that the reminder of the truth has any chance of being supplied."[17]

---

[15] Mill, John Stuart. *On liberty and Considerations on Representative Government*. Basil Blackwell. Oxford. 1946. p. 46.
[16] Ibid, p. 15.
[17] Ibid, p. 46.

However, Mill demands the expression of antithetical opinions even when these opinions are patently erroneous. He thinks that without debate and dispute with other opinions, our opinion (even if correct) would become degenerate and decadent. We would forget the arguments that brought us to believe in this opinion in the first place, and our opinion might turn into a sort of prejudice which we cannot anymore remember why we actually hold it. Through the confrontation with competing opinions our position is strengthened and our confidence is intensified. "Even if the received opinion be not only true, but the whole truth; unless it is suffered to be, and actually is, vigorously and earnestly contested, it will, by most of those who receive it, be held in the manner of a prejudice, with little comprehension of feeling of its rational grounds."[18] Without the clash with other opinions we might forget the original sense and meaning of our opinion, and our holding of it may stop generating any utility to us, or to our society. Mill's fourth argument goes even further, and warns us that the silencing of the opposite (even false) opinion endangers the common and prevailing idea, which we consider to be "the whole truth." He says that "the meaning of the doctrine itself will be in danger of being lost, or enfeebled, and deprived of its vital effect on the character and conduct: the dogma becoming a mere formal profession, inefficacious for good, but cumbering the ground, and preventing the growth of any real and heartfelt conviction, from reason or personal experience."[19]

It appears that examining the general question of freedom of expression with regard to the question of allowing Prof. Van Creveld the expression of his unusual (perhaps even bizarre) opinions in his academic course, would result in the conclusion that it might be more beneficial to allow the expression of these opinions, than trying to silence them. Here I think the claim which Katharine Gelber ascribes to the defenders of free speech that "hate speech may productively be countered through the maintenance of principles ensuring the freest speech possible, for as many people as possible,"[20] is quite correct. Leon Friedman expresses this attitude by saying, "if we are involved in public discussion of even the most hateful kind, the way to deal with it is by more speech".[21] This attitude is basically the argument in favor of tolerance, as Glen Tinder, following John Locke, explains it: "Belief cannot be determined by force. One can be compelled to act in a certain way but not to think in a certain way. It follows that intolerance, whether or not it is evil, it altogether futile."[22] The demand for principles that maximize free speech is probably the most efficient way of reducing hate speech, even though they have

---

[18] Ibid, p. 46.
[19] Ibid, pp. 46–47.
[20] Gelber Katherine. ibid, p. 2.
[21] Friedman Leon. "Freedom of Speech: Should It Be Available to Pornographers, Nazis, and the Klan." In: Freedman Monroe H. and Freedman Eric M. (eds.) *Group Defamation and Freedom of Speech.* Greenwood Press. Wesport. 1995. p. 317.
[22] Tinder, Glen. *Tolerance and Community.* University of Missouri Press. Columbia. 1995. p. 22.

their problems too. Frederic Schauer, who deals a lot with the principle of free speech, says that when such a principle is accepted "there is a principle according to which speech is less subject to regulation (within a political theory) than other forms of conduct having the same or equivalent effects."[23] And though he admits that speech, being not a self-regarding act, can, and frequently does cause harm,[24] it should be genuinely protected. He stresses that the fact that free speech is 'other regarding' is the main reason for protecting it.[25] Thus, even if Prof. Van Creveld's course can be considered as defamation, or even obscenity, we should still allow it. The most we could expect is self-censorship by the professor, but we should not use censorship methods to fight against undesired opinions. We should allow the expression of these opinions and make efforts to refute them, in order to convince other people not to accept them, or to accept their polar opposites.

However, we still can ask whether it is appropriate to express such opinions under academic cover where they are viewed as the opinions of an expert, and not merely the opinions of an eccentric. In academia these claims may appear as scientific facts or truths, and may receive stronger validity or more obligatory or persuasive status to the students in the class. At this point I want to explain briefly why academic freedom is so important. We tend to think that we have to maximize freedom and offer a multiplicity of lifestyles and viewpoints if we expect academics to be "original" or "creative."[26] Following John Stuart Mill, we believe that "There is always need of persons not only to discover new truths, and point out when what were once truths are true no longer, but also to commence new practices, and set the examples of more enlightened conduct, and better taste and sense in human life."[27] These achievements and tasks are entrusted to only a very specific and limited circle of people, and in the current era, many of them are active in academic institutions. It has become part of the academic mission to make social and scientific progress, and this progress results from the work of these savants. These people are very few, but Mill thinks that without them, "human life would become a stagnant pool."[28] And for being able to enjoy the fruits of the genius and work, Mill argues, "it is necessary to preserve the soil in which they grow. Genius can only breathe freely in an *atmosphere* of freedom."[29] Thus, we should preserve the two necessary elements for human development: freedom and an array of viewpoints and approaches, at least in academic institutions.

---

[23] Schauer Frederic. *Free Speech: A philosophical Inquiry*. Cambridge University press. Cambridge. 1982. p. 7.
[24] Ibid, p. 10.
[25] Ibid, p. 11.
[26] John Stuart Mill quotes Wilhelm Von Humboldt, who thinks that these are necessary for originality. See ibid, pp. 50–51.
[27] Ibid, p. 57.
[28] Ibid.
[29] Ibid. The emphasis is in the original.

Rodney Smolla examines the issue of academic freedom and says, "Principles of free speech and academic freedom should certainly be understood to give faculty a large measure of independence of how they present materials in class on matters relating to race, sex, or sexual orientation. Viewpoint discrimination should not be permitted, even when the university regards the view espoused by the professor as repugnant. Thus a professor should have the right to espouse bona fide academic opinions concerning racial characteristics or capabilities, even though most people of good will and good sense on the campus would find the opinions loathsome."[30] However, Smolla insists that professors should refrain from hate speech. Those who think that hate speech should not be allowed within universities argue, "Implicit in the University's core mission is the regulation of expression to enhance its quality."[31] Similarly, the proponents of campus restriction on hate speech maintain, "official toleration of racist insult interferes with the educational process. Hate speech hinders learning and class participation, undermines self-confidence and alienates a victimized student from the rest of the student body. As a result, hate speech adversely affects both academic performance and social integration within the university community".[32]

I argue that even though Prof. Van Creveld's course includes hate speech, it should not be censored by the university for purely academic reasons. I follow Wojciech Sadurski's approach, which considers that "universities, as places primarily devoted to learning and research, should be prepared to tolerate more speech than should the global society. This is because "the search for truth" in more dominant in universities as an aim of collective enterprise than in society at large; because the universities' mission is to explore a wider range of ideas in a more thorough way than in public discourse outside, for example in the mass media; because universities are laboratories of collective life where experimentation is more vital than elsewhere, and experimentation demands more tolerance of expression which is seen as harmful by many."[33]

In my opinion, academic freedom requires that we allow Prof. Van Creveld to teach whatever he wants, even when his opinions irritate most of the students and faculty. Beyond the possibility that he may illuminate our minds with exciting scientific discoveries or truths, to which we were never introduced or exposed before, we should allow him to express and teach his opinions even if we think that they are vacuous, shallow, or even morally wrong. We should remember the value of variety, and that "it is good that there should be differences, even though not for the better, even though... some should be for the worse,"[34] as Mill reminds us. We just have to remember that academic freedom also requires not

---

[30] Smolla Rodney A. *Free Speech in an Open Society.* Alfred A Knopf, Inc. New York 1992. p. 213.

[31] Byrne J. Peter. "Racial Insult and Free Speech Within the University." Georgetown Law Journal. 79. (1991). P. 416. Quote is taken from: Sadurski Wojciech. *Freedom of Speech and Its Limits.* Kluwer Academic Publishers. Dordrecht. 1999. p. 184.

[32] Sadurski Wojciech ibid, p. 184. There he quotes Lawrence.

[33] Ibid, pp. 184–185.

[34] Mill, ibid, p. 66.

coercing students to hear opinions that irritate them. Thus, we must ensure that this controversial course will not be compulsory for any student.

However, the expression of such opinions as those of Prof. Van Creveld has additional significance as related to the very nature of academic work and research. And for these reasons I would allow him to teach that course even if it involves hate speech. The assessment of the research and academic abilities of professors is a long and complicated process of monitoring their work throughout many years and once they receive the official stamp of approval by the university as a professor, they automatically command respect based on the title. Hence, we have to be sure that the professor actually deserves this authority and respect.

If we cannot monitor this professor by, inter alia, listening to his lectures or reading his researches and articles, we will not be able to assess his scientific abilities and depth of his research, and the extent to which we should take his current opinions and ideas seriously. Without the expression and publication of his opinions, we would not be able to know, whether he thinks, for example, that the mental or intellectual differences between genders result from physical or biological differences, or whether he thinks that the low representation of women in certain domains has something to do with their social status throughout history. All this information can be clarified only if we allow that professor a stage where he can express his ideas.

And then, after we have clarified all the above issues, we must reassess the professor's research abilities, the depth of his arguments, his methodological competence, and some other components, which give the honored professor his current status and academic authority. Maybe after listening carefully to his claims and arguments we will be convinced, and thus, compelled to change our opinions and positions about many issues, including our ideas about the equality between genders. On the other hand, we might be compelled to reassess the professor's status and his research in other domains. After all, in the era of specialization we cannot always thoroughly understand studies conducted in different fields. We tend to assume that if we do not understand, realize, or agree with the expert's position, we should accept it, since he/she has the authority and say in the domains, to which he/she is supposed to be an expert. When such an expert expresses opinions, which we can understand, evaluate, and assess, it might give us a clue about the quality of the rest of his/her work. The probability that we can understand or assess theories about equality between genders is higher than that of theories in military history. We assume that theories about equality are, at least to some extent, ideologically biased, and affected by the researcher's worldview. Thus we usually assess these theories with regard to their social and academic context. However, we can consider racist or sexist remarks by faculty members in the course of their teaching "as possible evidence of the low intellectual qualifications of the lecturer."[35] For me this is a very good reason to allow the above course, since it

---

[35] Sadurski Wojciech. Ibid, p. 187.

may give us relevant information about the lecturer, which we could not attain in any other way.

In our case, in order to reassess our views or form an opinion about equality between genders, or to assess the professor's academic competence, we have to allow him the expression of his opinions and ideas, at least in front of those who want to listen to them. We might even be rewarded with the ability to understand that theories in the domain of military history are also ideologically biased, and affected by the researcher's worldview. We may suspect that our disagreement with the expert about issues from domains of his expertise is not a result of misunderstanding but of ideological dispute, or of a different worldview. In order to open the gate to such opportunities, we should allow professors to express their ideas, and it is our mission to criticize and evaluate them. It is also our duty to deduce comprehensive conclusions from this evaluation, including the reassessment of the academic and professional status of these professors.

To conclude this case, and the issue of hate speech, not only in academic institutions but as a whole, I think that Leon Freedman was correct in saying, "We must tolerate the speech and deal with it in the only forum available to us, namely in the field of public debate ... It is smoothing and reassuring because it frees us to express our political arguments in the most forcefully way we want, leading to 'uninhibited, robust, and wide-open' debate, and requiring us to listen and deal with the political speech of others."[36] The way to fight hate speech should be by arguments and debates, and not by using censorship. What we should do is, "Let truth and falsehood grapple. Who ever saw truth bested in a free and open encounter?"[37]

---

[36] Friedman Leon. Ibid, p. 316.
[37] Ibid, p. 317.

SECTION B. MEDICAL ETHICS

Of the myriad of ethical debates going on in the present-day public discourse, it would not be an exaggeration to say that medical ethics raises the most intensive and vehement arguments. The rapid progress in technology and bio-technology has far outstripped parallel progress, if there be any, in either moral or legal studies. It seems that by the time that ethical discussion, buttressed by legal considerations, grasps and is able to deal with a medical issue—one that is usually on the frontier of scientific discovery—technological progress has already flung us forward into more complicated or acute issues. This section deals with some of these issues.

The first chapter deals with the somber question of mercy death or mercy killing. Technological advances for extending human life and maintaining acutely-ill persons on life support may be perceived as both a blessing and a curse. Alongside new hope that is sometimes acquired by an extension of one's life-span, the preservation of a life gone awry often involves suffering for the patients and additional burdens for their caretakers. This chapter inquires, generally, into the legitimacy of both coercive treatment and abstention of such, for those who either refuse or cannot express their consent to medical care. A specific issue arising here, under the same theoretical umbrella, is the legitimacy of separating Siamese twins, when it is clear that at least one of them will die as a result.

The second chapter deals with the more optimistic but no less sensitive and complicated issue of organ donation or organ sale. The possibility of saving life or enhancing its quality by organ transplantation, and the unwillingness of (sufficient) people to donate the required organs, raises a question about the morality of selling human parts. Two difficulties are immediately encountered: First, it is clear that poverty-stricken people are the ones most liable to risk their own health in order to procure funds. The problematics of commerce in human organs is revealed in the current chapter, most notably exemplified by kidney "donations." Secondly, and perhaps more marked by philosophical principles, the value-laden question of the reification and marketability of the human body is analyzed through the paradigm of ovum-contributions. The first is a problem of justice; the second—one of values.

The third chapter deals with genetic engineering and reproduction. The field of genetic engineering encompasses the (insurmountable?) gap between technological progress and the inability of the ethical dictionary to respond to issues that arise in its wake. The most far-reaching debates in this area have to do with human reproductive cloning and stem-cell research. The conflicts between scientific demands and political or philosophical misgivings and qualms, sometimes make the issue intractable. We try to offer guidelines for therapeutic stem-cell research while, at the same time, banning human reproductive cloning, thereby attempting to address both sides of the dilemma consistently. A related issue, also discussed in this chapter, is the question of creating new siblings for the purpose of using their organs (usually bone-marrow) to medically assist their brothers and sisters. It is clear that future scientific/technological developments may, nevertheless, obligate us to rethink our own judgments—in this, as in all other questions in this section.

Much of this chapter deals with the danger of leaving such acute issues exclusively in the hands of the scientific community. This danger was clearly raised by Jurgen Habermas, whose book *The Future of Human Nature* discusses this in detail when he believes that the new technologies make a public discussion on the appropriate understanding of cultural form of life in general an acute matter. He believes that the members in the philosophical community no longer have good excuses for leaving such a dispute to biologists, engineers or businessmen. Genetic engineering challenges some of our most fundamental beliefs about morality. It enables us to control the physical basis which we are by nature, and as Habermas describes this problem, things that for Kant belong to the "kingdom of necessity," in the perspective of evolutionary theory, now become a "kingdom of contingency." Genetic engineering is now shifting the line between the natural basis we cannot avoid and the "kingdom of ends." This extension of control of our "inner" nature is distinguished from similar expansions of our scope of options by the fact that it changes what Habermas considers as the overall structure of our moral experience.

One of the most fundamental changes of this kind, as Habermas understands it, may be the uprooting of the categorical distinction between the objective and the subjective, and this dedifferentiation of fundamental categorical distinctions, which we have as yet, in the description we give of ourselves, assumed to be unchangeable. This differentiation might dramatically change our ethical self-understanding as the authors of our own lives and as equal members of our moral community.

However, genetic engineering might also have very promising consequences, particularly in the domain of stem-cell research. I will bring Gordon Graham's argument that Stem-cells can be used to repair organic damage, or to recreate diseased or malfunctioning parts of the human body. As such they present us with promising new therapeutic possibilities, some of which have already been extraordinarily demonstrated. The most familiar of these is bone marrow transplants in patients with leukemia, which can regenerate a healthy blood system. This chapter deals with the dilemmas that are raised from these possibilities but its main theme is that we should restrict the research to therapeutic cloning and confine the immense potential of stem-cell research only to negative eugenics—all the while maintaining tight control over the researchers and scientists involved. Although this supervision should be, primarily, the mandate of the scientific community, the international community and society as a whole should share the burden of monitoring the scientists. All of us should ensure that scientists do not cross the border between negative eugenics, which prevents diseases, and positive eugenics, which might genetically enhance the species which we know as homo sapiens, but with genetic enhancement might be changed into something else.

Chapter 5

MERCY DEATH OR KILLING

One of the most acute issues in current medical ethics is the problem of euthanasia. This term originally comes from the Greek word that originally meant "a good death" (*eu*—well, *Thanatos*—death). Needless to say that in every discussion about life and death we hold the a priori assumption that "life is good and that existing life should be preserved as a matter of course, unless some overriding principle supersedes the innate value of an ongoing life."[1] This assumption is one of the reasons for the immense difficulties of the discussion about "good death." The issue of euthanasia became extremely complicated during the 20th century due to rapid technological progress that enabled maintaining the lives of terminal patients, even unconscious terminal patients, for extended periods of time. However, while we may possess powerful life-prolonging medical technology, on the ethical level "we are unable to find meaning in death or to bring our lives to a meaningful close."[2] Thus, death as a whole, and euthanasia in particular, have become complex and painful issues in modern society.

The discussion in the current chapter deals with three different meanings that are usually accorded to the term "euthanasia." However, all three forms of euthanasia assume that the bedridden person under discussion has an acute, terminal, chronic illness that medical treatment cannot hope to cure or even ameliorate. In such a situation we tend to believe that if something "is taken from a dying person, it is nothing he wants to keep and the act is one of giving rather than taking."[3]

---

[1] Whiting Raymond. *A Natural Right to Die.* Greenwood Press, Westport, Connecticut, p. 171. The emphasis is in the original.

[2] Hardwig John. "Dying at the Right Time: Reflections on (Un) Assisted Suicide." In: LaFollette Hugh. (ed.), *Ethics in Practice*. Blackwell Publishers, Cambridge, Massachusetts, 1997, p. 64.

[3] Barrington, Mary Ross. "The Case for Rational Suicide." In: Downing A. B. and Smoker Barbara (eds.), *Voluntary Euthanasia*. Peter Owen Publishers, London, 1969, p. 247.

The first type of euthanasia is usually called "allowing someone to die." This type refers to the desire of terminally ill persons, who are no longer helped by medical treatments, to be allowed to die naturally, in peace and dignity, rather than be kept on life support. The crucial distinction of this form of euthanasia is that no active termination of life is carried out, but only an abstention from additional medical treatment that only lengthens patients' lives but does not improve their health condition or quality of life. The main point of this type of euthanasia is not to artificially lengthen life through modern technology when technology cannot cure the patients or improve their condition. Instead, the patient only receives drugs or treatments to relieve pain or discomfort.[4]

The second type is called "mercy death" and refers to direct actions to terminate the lives of terminally ill patients that explicitly request to die. We can consider this a kind of assisted suicide, but not in the sense of "preemptive suicide." The usual term of "preemptive suicide," as C. G. Prado uses it, does not refer to cases of "escaping actual, intolerable circumstances, but avoiding foreseen demeaning decline and needless suffering."[5] I use mercy death to refer to terminally ill persons who suffer and cannot put an end to their lives by themselves and so they must ask someone else to assist them to die, usually by painless means or methods. Here the patients make an autonomous decision to choose death over their currently painful lives, and need someone else to help them carry out their own desires.[6] An equivalent term to mercy death is *voluntary active euthanasia* (VAU), or euthanasia at the request of the patient.[7] Those who support this doctrine believe that just as "a free and autonomous person can renounce and relinquish any right, *provided only that his choice is fully informed, well considered, and unforced,* that is to say, *fully voluntary*," [8] one can relinquish his right to life, and have voluntary active euthanasia.

The third type of euthanasia is also a form of "mercy killing" that takes direct actions to terminate a patient's life. However, unlike the previous type, here it is carried out not only without the patient's explicit request, but even without the patient's consent. In this type, the decision that the patient's life is no longer meaningful is taken by someone else, assuming that if the patient could express his/her desire or will, he/she would ask others to put an end to his/her life. This type of mercy killing is also called *non-voluntary active euthanasia* (NVAU), which refers to cases where "euthanasia preformed on those who do not have the mental ability to request euthanasia (such as babies or adults with advanced

---

[4] See Thiroux Jacques. *Ethics: Theory and Practice*, Sixth Edition. Prentice Hall, Upper Saddle River, NJ, 1995. p. 213.
[5] Prado, C. G. *The Last Choice.* Greenwood Press, Westport, Connecticut, 1998, p. 2.
[6] See Thiroux Jacques. Ibid, p. 213.
[7] See Keown John. *Euthanasia, Ethics and Public Policy*. Cambridge University Press, 2002. p. 9.
[8] Feinberg Joel. *Rights, Justice, and the Bounds of Liberty*. Princeton University Press, Princeton, 1980, p. 250.

dementia) or those who, though competent, are not given the opportunity to consent to it."[9]

The fundamental assumption of all three types of euthanasia is that since terminally ill patients bear the suffering involved in staying alive, they have absolute rights over their bodies which are their ultimate and exclusive prerogatives. This assumption is widely accepted and establishes a comprehensive acknowledgement, if not a consensus, of the legitimacy of the first type, namely that of allowing someone to die. Once a patient is fully aware of his/her health condition, he/she has the ultimate prerogative to decide which treatments to receive or reject. When such a patient refuses a new treatment or asks to stop an ongoing treatment, this request should be (and usually is) respected. The principal justification for this is the desire to reduce unnecessary pain or suffering from the patient. Since the patients are the ones who suffer, it is their right to demand the end of this suffering and it is society's mission to comply with their request (usually but not necessarily via the doctors). Sissela Bock explains the rationale behind this demand by saying: "The greater the power people have to make choices in their own life, the more reasonable it is for them to seek added control in warding off suffering at the end of life."[10] If such a decision also brings relief to the patient's relatives, this is to be considered only as an additional benefit since the patient's decision must be fully autonomous and not determined by the desires of the relatives. While the patient may take the desires of relatives into account, it is his or her decision to make such considerations. The medical personnel should definitely remain indifferent to the preferences or desires of the relatives.

The absolute right of a patient over his/her body also dictates general public opinion about the third type of euthanasia: mercy killing. In cases where patients did not express an explicit desire or request to terminate their lives (even when the patient is physically unable to express their desires, for example due to a persistent vegetative state, irreversible coma or even brain death), medical personnel usually abstain from any action that may hasten the patient's death. The rationale behind this practice is the belief that it is the patient's ultimate right to decide whether life is meaningful, and whether and when is the time to terminate it. When the patient cannot evaluate his/her life and make an explicit request to terminate it, neither the medical personnel nor the family are allowed to take such a decision regarding the patient's life. All involved parties must make the highest possible efforts to maintain the patient's life, even if only for the sake of the minor chance that the patient would accord some value to his/her life even under the current circumstances.

The second form—mercy death—is much more complicated especially regarding those cases where the patients are so sick that they cannot terminate their

---

[9] Keown John. Ibid, p. 9.
[10] Bock Sissela. "Choosing Death and Taking Life." In: Dworkin Gerald, Frey R. G and Bock Sissela (ed.), *Euthanasia and Physician-Assisted Suicide*. Cambridge University Press, 1998, p. 84.

lives by themselves but require the active intervention of someone else. Although the patients possess the ultimate right to decide whether there is still value to their lives, and if and when they should be terminated, no one (not even the medical personnel) are required or duty-bound to assist them. This may lead to a situation where although a patient wants to die, he is unable to do it by himself and none of the people around him will comply with his request to assist him to die.

However, when it is obvious that the terminally ill patient is of sound mind and sincere desire to die, we should fulfill her desire to put an end to her misery even though no one can force us to do so. If we can find someone who agrees to carry out a concrete action to hasten the patient's death, society should allow this and enable the patient to die in dignity and peace. This may show society's respect to the patient's ultimate right over her body, and society's obligation to reduce suffering within it.

This approach usually raises two difficulties. The first is the "slippery slope" argument that holds that allowing mercy deaths may lead to the termination of lives of patients who are not terminal, or who still have a desire to continue their life. The second difficulty may be the harm of what Ronald Dworkin calls "the sacredness of life."

With regard to the slippery slope argument, R. G. Frey has made two essential clarifications that are worth mentioning here. "First, the argument is about the likelihood of disastrous slope consequences coming to pass: it is not argument involving causal necessity."[11] Frey reminds us that this argument does not say that we shall be compelled through causal necessity to descend the slope, but that it becomes empirically very likely that we would. The second clarification Frey makes is that since the slippery slope argument is not a causal argument, it is neither a quasi-logical one. "That is, the thought is not that one is inexorably led down the slope, say, of taking life as a result of line-drawing or boundary disputes, however real those disputes may at times be."[12] In any event, the fact that empirical means may lead us down the slope of taking life, requires that we should use safeguards and safety measures to reduce that fear.

Thus, in order to avoid the slippery slope argument, we must require the stringent fulfillment of at least two conditions: the objective one (which requires that the patient is really and undoubtedly terminally ill), and the subjective one (which requires that mercy death is the patient's autonomous, sincere and conscious desire). A significant term in the context of this discussion is informed consent of the patient.Dan Broke elaborates on this and holds that sound health care decision-making involves two components that should be taken into account. The first is the "objective" one, which relates to "empirical facts about the nature of the expected outcomes for the patient of different treatments (including the

---

[11] Frey R. G. "The Fear of a Slippery Slope." In: Dworkin Gerald, Frey R. G., and Bock Sessila (ed.), *Euthanasia and Physician-Assisted Suicide.* Cambridge University Press, 1998, p. 44–45.
[12] Ibid, p. 45.

alternative of no treatment). These include a determination of the patient's physical status, including his or her prognosis and expected condition under alternative treatments."[13] These are usually facts and assessments, which will be best judged by the physicians, but even among different experts there can be disagreements about both facts and assessments of the same patients (so the term "objective" is somewhat misleading). In any event, the patient should have maximal information about his health condition including the disputes among the experts regarding both diagnosis and prognosis, before he makes any decision about the treatments offered to him. This component relates to the "informed" part of the doctrine. The second component is the "subjective" one, and this component relates exclusively to the patient's considerations. This "subjective" component "involves assessing the relative importance or value to this particular patient of the features and consequences of the alternative treatment outcomes". The evaluation of the outcomes, and of their effects on the patient's well being, is a subjective component in the sense that it depends on the patient's goals and values. These values and goals may be quite different from the physician's, or even from most persons, depending as they do on the patient's preferences, abilities and opportunities.[14] This component establishes the patient's possibility of expressing free and autonomous consent. When a patient expresses a preference that is based on his best judgment and authentic belief, that is, his decision is well-informed and autonomous, society has to respect the patient's dignity by accepting and respecting his decision.

The second difficulty might even strengthen the demand to permit mercy death from a secular point of view. This is because the secular world-view holds that part of the sacredness of human life or the intrinsic value of life, results from the subject's evaluation of her own life as meaningful and worthy of living. When a patient's life is saturated with suffering and misery, and all her hopes and desires are to put an end to such a life, we should consider her as a human being who has free will, and assist her to fulfill her own will or desire. By doing this we show respect to her very humanity that confers intrinsic value to her life. Banning mercy death would impose the responsibility for the patient's future suffering and for the harming of the patient's subjectivity, on society as a whole. There is a strong tendency to believe that "a man suffering with severe, intractable pain due to an incurable illness and whose death is inevitable and imminent,"[15] has the moral right to voluntary euthanasia. That is, this person has a right "to be able to die with dignity at a moment when life is devoid of it."[16] Elisabeth Kubler-Ross

---

[13] Brock Dan W. *Life and Death*. Cambridge University Press, Cambridge, 1993, p. 27.
[14] Ibid, p. 28.
[15] See Ayd Frank J. Jr. "Voluntary Euthanasia: The Right to Be Killed". In: Ostheimer Nancy C. and Ostheimer John M. (eds.), *Life Or Death—Who Controls?* Springer Publishing Company, New York, 1976, p. 238.
[16] Mannes Maria. "The Good Death." In: Ostheimer Nancy C. and Ostheimer John M. (eds.), *Life Or Death—Who Controls?* Springer Publishing Company, New York, 1976, p. 225.

describes this by saying, "When a patient has reached the stage of acceptance and the family is also at peace, the patient often asks to stop all life-prolonging procedures. We would respect this request under most circumstances, especially if we are sure that the patient has no chance of cure or a remission."[17]

I would like to clarify the distinction between active and passive euthanasia, as this distinction is crucial for medical ethics. In the terminology I use, passive euthanasia refers to allowing someone to die while active euthanasia consists of both mercy death and mercy killing. For those who consider this distinction as relevant for the discussion on euthanasia, "it is permissible, at least in some cases, to withhold treatment and allow a patient to die, but it is never permissible to take direct action to kill a patient."[18] This doctrine was widely accepted by doctors and in 1973, was even endorsed by the American Medical Association (AMA) in the first policy statement on euthanasia ever issued by this association. This statement says:

> The intentional termination of the life of one human being by another—mercy killing—is contrary to that for which the medical profession stands and is contrary to the policy of the American Medical Association.
>
> "The cessation of the employment of extraordinary means to prolong the life of the body when there is an irrefutable evidence that biological death is imminent is the decision of the patient and/or his immediate family. The advice and judgment of the physician should be freely available to the patient and/or his immediate family."[19]

Following James Rachels, I consider this distinction to be impertinent, at least with regard to the question of the legitimacy of any sort of euthanasia. Rachels argues that practically, there is no real difference between cessation of treatment, which will result in the patient's death, and "the intentional termination of the life of one human being by another."[20] Rachels believes "that active euthanasia is no worse than passive euthanasia and, therefore, that the conventional doctrine is false."[21] He examines the main argument in favor of the conventional doctrine, which maintains that in passive euthanasia the doctor does not do anything to bring about the patient's death, and the patient dies from the disease from which he is suffering. In active euthanasia the doctor is the one who kills the patient. Rachels argues that when the doctor lets the patient die, he still does something, and from the moral point of view one can perform an action by way of not performing any action, and thus is morally responsible for the consequences of that event.[22] Thus,

---

[17] Kubler-Ross Elisabeth. "Prolongation of Life." In: Ostheimer Nancy C. and Ostheimer John M. (eds.), *Life Or Death—Who Controls?* Springer Publishing Company, New York, 1976, p. 224.
[18] See Rachel James. *Can Ethics Provide Answers*. Rowman & Littlefield, Lanham, 1997, p. 63.
[19] Quotation is taken from Rachel James. Ibid.
[20] Ibid, pp. 66–67.
[21] Ibid, pp. 67.
[22] Ibid, pp. 67–68.

when one opposes active euthanasia for moral, religious or personal reasons, this person should also oppose letting someone die, for the same reasons. And if one condones letting a terminally ill person die, for mercy reasons, there are many circumstances that the same reasons would lead him to support active euthanasia. Thus, I agree with Rachels, and do not deal with the conventional distinction between active and passive euthanasia. I consider both to be "acts directed toward bringing about a merciful, peaceful and dignified end."[23] Accordingly, I use a different distinction: allowing someone to die refers to cases where there is no external intervention in the patient's natural death; and mercy death or mercy killing both involve direct action (or direct omission) to terminate the patient's life (though they are apparently different from each other).

In the coming paragraphs I discuss two different cases; the first involves allowing someone to die, and the second involves mercy killing. I will try to present the complexity of such cases, and the sensitivity which we should show when we deal with such painful issues.

CASE 1

In September 2000 Judge Yaffa Hecht, a Jerusalem District Court Judge, ruled to allow a 76-year-old woman to discontinue dialysis treatment, to comply with the patient's explicit request. This decision would be valid only if the woman would maintain her refusal after additional attempts to convince her. Two days later, the woman died in the Hadassah Ein Kerem Hospital in Jerusalem. It is worthy of noting that this woman had tried to refuse intrusive medical treatment previously when the hospital's Ethics Committee forced her to have heart catheterization against her explicit wishes. Another relevant point for this discussion is that a psychiatrist evaluated her mental state and decided that her mental faculties were intact.

The doctor who asked the committee's permission to perform the catheterization against the woman's objections, argued that the woman's objections were due to personal fears and not as a result of religious or conscientious reasons. He said that his impressions were that the woman did not despise her life. He estimated the she had a greater than 80% chance of death by refusing catheterization while the procedure would reduces the risk to her life to 40 or 50%. The hospital's ethics committee accepted his arguments and ruled to enforce the catheterization, due to what the committee considered the woman's acute health condition and serious risk to her life. In the second case, the members of the Ethics Committee said that they would force the woman to undergo dialysis even against her will as her life was in danger and she seemed to have a desire to go on living.

---

[23] Hyde Michael J. *The Call of Conscience.* University of South California Press, 2001, p. 16.

This contention of the Ethics Committee members is infuriating: even a threat to the woman's life does not nullify the woman's absolute and exclusive right over her own body. We must freely acknowledge the woman to be an adult who is responsible for her decisions and desires; her body is not property of the hospital or its medical staff. Just as we do not use the "life threat" rationale to prevent sportsmen from engaging in dangerous sports activities or professionals from working in dangerous professions, so we should not take advantage of a patient's helplessness to force her to undergo life-saving procedures. Even the medical authorities must respect the fact that the patient is a free and rational agent and thus respect her desires even when they involve a threat to her life. The woman's ultimate and exclusive right over her body and life, bestows her with the authority to decide which risks she wants to take, and at what price. We should remember that part of the intrinsic value of her life as a person, which Ronald Dworkin calls *The Sacred*, has been created "through the choices she has made, her own creation."[24] Dworkin cites the two Greek terms for life: *zoe,* which relates to physical life, and *bios* which means "a life as lived, as made up of the actions, decisions, motives, and events that compose what we now call a biography."[25] The second term, according to Dworkin, establishes the sacred value of human life, which gives it an intrinsic value. If we ignore the woman's desires, preferences and will, we show disregard to an essential component of the sanctity of her life, and thus we do not respect her status as a human being whose intrinsic value should be inviolable. In other words, we harm the sacredness of her very humanity.

There is another important issue regarding the impression of the Ethics Committee that the woman did not want to die. This means that they give more weight and importance to their own impressions of the woman's mental state, than to her explicit request. This is not legitimate: they must understand that they have to comply with the patient's wishes even when these oppose their own conceptions or preferences. They cannot be arbitrary about choosing when to comply with a patient's requests and when to ignore them. Thus they must always respect the patient's desires to refuse medical treatment even when doing so may cause the patient's death.

Another issue is the doctor's argument that the patient's refusal was based on "fears," and not on religious or conscientious reasons and considerations. Firstly, we do not accord more importance or weight to religious or conscientious considerations over a patient's preferences for any other reason. Thus, fear is a legitimate reason unless it harms the patient's reasoning or judgment abilities. A patient who understands the risks of certain medical or surgical procedures may well prefer a shorter life with less pain to a longer but more painful life. We should remember that the patient's absolute right over her life includes her exclusive right to evaluate her own life, and to decide for herself to what extent

---

[24] Dworkin Ronald. *Life's Dominion*. Harper Collins Publishers, 1993. p. 82.
[25] Ibid, p. 83.

Mercy death or killing

her life is still appropriate and valuable. Once she makes her own decision and explicitly expresses her will and desire, society should respect her decision even if it differs from her own. Thus, we should restrict the authority of the Ethics Committee only to those cases where the patient did not express any preference or will, or to cases where there are doubts about her mental judgment—not to cases where we know that her clear and lucid preferences are different from ours.

To sum up this case I want to stress that the decision about allowing someone to die should remain the patient's absolute and exclusive right. Just as we cannot cause someone to die, we cannot force one to keep on living against one's desire and will. After all, it is the patient herself who bears the cost of any decision regarding her life, and thus she should have the final word. In cases of terminally ill patients, society should not "err in minimizing the evil of human suffering and overrating the value of merely biological life in the absence of a human person, or in the presence of a human person whose suffering are too severe for him to have a human life, even though his heart beats on,"[26] as Joel Feinberg reminds us. Here I adopt the doctrine, described by Carl Wellman, which maintains, "The right to life is an option right that, paradoxical as it may seen, includes the patient's right to die."[27] However, there is no duty on any particular agent to assist the dying patient, but definitely one who wants or agrees to do so should be allowed to. The doctors' status and authority here should remain voluntary. Here I accept A. D. Woosley who discusses this issue and asks:

> If patients ... are to have the right to have life support removed, what of the physicians who are to terminate, or to give the order to terminate? Clearly they should not be required to terminate it, if that is contrary to their best professional and moral judgment ... But if in the physicians' judgment termination would be best, and if in their judgment the person consenting to termination has authority to do so and really appreciates the alternatives between which he is choosing, then they should terminate.[28]

The next case goes even further and discusses the issue of mercy killing under very specific circumstances that might be acceptable even to those people who generally object to mercy killing.

CASE 2

In August 2000 British Judge Robert Johnson ruled that 2-week-old Siamese twins—under the pseudonyms Jodie and Mary—should be separated from each

---

[26] Feinberg Joel. *Freedom and Fulfillment*. Princeton University Press, 1992, p. 282.
[27] Wellman Carl. *Real Rights*. Oxford University Press, Oxford, 1995, p. 5.
[28] Woozley A. D. "Euthanasia and the Principle of Harm." In: Rachels James (ed.), *Moral Problems*, Third Edition. Harper and Row Publishers, New York, 1979, p. 504.

other, even though one of the baby girls would definitely die during that process. All the doctors supported this procedure but the babies' parents opposed it. As devoutly religious people, they believed that their daughter's destiny should remain in God's hands and they could not agree to a procedure that would result in the death of one of their children to enable the survival of the other. They raised the following question: If everyone has a right to live, how they sanction the killing of one of their daughters to enable the survival of the other, when both babies have the same right to life?

Mary and Jodie were born joined at their lower abdomen, on August 8 at St. Mary's Hospital in Manchester, Northern England. Of the two, Jodie was more vital and lively and Mary was suspected of suffering from brain damage. Mary absolutely relied on Jodie for functions of the heart and lungs that were physically located in Jodie's body, and Mary only lived because she was attached to her sister's body. There was a consensus among the doctors that if the twins would not be separated, they both would die within the next three to six months. However, the doctors were also certain that once the twins would be separated, Mary would immediately die. The surgery also might leave Jodie severely handicapped, or even cause her death during the operation. However, if she would survive the operation she still had a good chance at a normal and healthy life.

Under these specific circumstances, the only possible way to separate the babies was to detach Mary's body, which did not have a heart and lungs of her own, from these organs located in Jodie's body. This would immediately terminate Mary's life. Justice Johnson, giving his ruling in public after a private hearing, said that separation of the babies would grant Jodie a chance or expectation of a normal life, while condemning Mary to immediate death.

Apparently this situation is similar in certain ways to the procedure of embryo dilution in cases of pregnancies with multiple embryos. In such cases, some of the embryos are destroyed in order to enable the remaining embryos a better chance of survival. However, the current case refers to babies and not embryos, rendering the problem much more complicated. While the rights of the fetus rights are debatable and unclear, the rights of babies are unambiguous and definite. Another aspect making the current issue more complicated is the parents' objection to any external intervention in the babies' life. So the decision of the judge to terminate Mary's life was a complex one.

First, the judge had to pinpoint the morally relevant considerations, and decide how much weight he should ascribe to each consideration. For example: does the fact that Mary's health condition is much worse than that of Jodie, and she might even be blind, reduce the extent of her right to life? Does the fact that the common heart and lungs are located in Jodie's body increase her right to life, over that of Mary, or even constitutes Jodie as the exclusive owner of those limbs? My answer to both questions is negative. An inferior physical condition cannot harm the right to live, and the fact that the common limbs are located in Jodie's body cannot be understood as the elimination of Mary's rights over the common heart and lungs, or at least her right to use them. What is extremely relevant is the fact that in case of separation, the location of the common limbs makes possible only the detachment of Mary from the common limbs, even though she may possess equal rights over these common limbs.

In my opinion, the most critical question in this case is whether the court could decide on Mary's death, particularly when the parents were strongly against it. No doubt that the parents' opinion should have a certain weight in making such a decision, mainly since this decision means the killing of their daughter Mary. Moreover, they are the people who will raise the other daughter, Jodie, with all the difficulties involved (since there is a risk that she may suffer from health problems due to the surgery). However, whilst the parents sincerely represent their interests, and the interests of their daughter Mary, the court has to consider the interests of all parties involved, including Jodie.

In the above situation, the best-case scenario is that both twins would survive. The worse scenario is that both would die. If the court accepts the medical opinion that there is no way that both could survive together more than six months, it has to decide in favor of the less tragic option of allowing at least one of them to live, and not in favor of the worst alternative, that would condemn both to death. Such a decision would represent both Jodie's interest in staying alive as well as the public interest, which requires that society should do its best to sustain life and prevent unnecessary death.

However, what about the parents' demand to leave the decision about their daughters to God's will? Ignoring their demand is tantamount to disregarding their religious faith and beliefs, and even violating their freedom of conscience. Complying with their demand would be, according to the court's understanding of the situation, deserting or even sacrificing Jodie. In such a situation, the court, as a national and not religious institution, cannot stand aside and ignore medical opinion (even though it is more like a prognosis than a medical diagnosis). If it is the court's obligation to save life it has to fulfill this duty, even at the cost of shortening Mary's life by a couple of months in order to facilitate Jodie's long-term survival. In the current situation, the striving for the best—that is the survival of both twins, at least temporarily—would lead to the worst consequence in the long term: the death of both. What is left to the court is to allow the termination of Mary's life in order to save Jodie's, in spite of the parents' protests, being acutely aware of the pain involved in this decision—and not choose on a decision that relies on a miracle.

The problem of mercy killing is much more complicated in cases where the mercy killing of one person could save the life of another, when the first person could still survive for a while, albeit in a vegetative condition. Here, it is not so obvious that we can sacrifice one person's life in order to save another, for example by transplanting one or more of the person's organs. However, if Patient A is in a vegetative state with no hope of recovery and will *surely* die sooner or later, and Patient B will *surely* die without the immediate transplant and is likely to survive the surgery, then I think the same principle of saving human life may serve as a guideline. This is because we do not really kill the terminally ill person, but only hasten his imminent death in order to save another human life. Thus we should at least consider this option, even without the explicit consent of Patient A. In fact, every harvest of hearts and livers, and many harvests of lungs and kidneys involve the killing of the donor, since these organs are effective only if they are taken from a live body. The main difference from the Mary-Jodie instance is that if the patients or their families oppose the mercy killing of their relative it should not be done, even at the cost of another person's life.

People who absolutely object to any kind of euthanasia frequently use the religious argument which claims that life and death are to be decided solely and exclusively by God. Although I cannot contest this argument since it belongs to the realm of belief, I still refer to a more moderate version of such an approach as espoused by Ronald Dworkin who holds that the sacredness of human life results from *both* human and natural investments; therefore, even those who think that the natural investment in human life is dominant can understand that euthanasia does not necessarily frustrate nature. Dworkin says that "they may plausibly believe that prolonging the life of a patient who is riddled with disease or no longer conscious does nothing to help realize the natural wonder of a human life, that nature's purposes are not served when plastic, suction, and chemistry keep a heart beating in a lifeless, mindless body, a heart that nature, on its own, would have stilled."[29] Without taking this concept into the religious sphere, I just want to state that those who believe in God's compassion and mercy might agree that it is God's will that a hopeless patient will not have to continue suffering.

Those who believe in human investment in life can also insist, "Euthanasia sometimes *supports* this value."[30] I think Dworkin's argument is very convincing. He assumes that those who believe in the sanctity of human life, also believe that "once human life had begun it matters, intrinsically, that life goes well, that the investment it represents be realized rather than frustrated."[31] Someone who treats their life as something sacred thinks that it is intrinsically important that he lives with dignity and integrity. And when such a person thinks that his life would not be worth living if he were hooked up to a dozen machines, and this person went to the trouble to make arrangements in advance to avoid this situation, then society should respect his wish to die. In fact, allowing him to die is a way of according respect to the sanctity of his life. An honorable society would not coerce a patient or seek to impose a collective spiritual judgment on everyone. An honorable society should "allow and ask its citizens to make the most central, personality-defining judgments about their own life for themselves."[32] This policy

---

[29] Dworkin Ronald. Ibid, p. 215.
[30] Ibid. The emphasis is in the original.
[31] Ibid, p. 215.
[32] Ibid, p. 216.

Mercy death or killing

will show respect to what Bertram and Elsie Bandman believe: "If there are any moral rights at all, there is at least one prior right founded on justice—the equal right of all persons to be free to decide to live or die."[33]

It is about time for modern society to legitimize people's decisions that their lives "shouldn't be lived, aren't worth living, at least not to their end."[34] Giving weight to the human investment in people's lives may lead us to follow Joseph Fletcher who believes that "It is harder morally to justify letting somebody die a slow and ugly death, dehumanized, than it is to justify helping him to escape from such misery. This is the case at least in any code of ethics which is humanistic or personalistic, i.e. in any code of ethics which has a value system that puts humanness and personal integrity above biological life and function."[35]

---

[33] Bandmann Bertram and Elsie. "Rights, Justice and Euthanasia." In: Velasquez Manuel and Rostankowski Cynthia (ed.), *Ethic: Theory and Practice*. Prentice Hall, Inc. Englewood Cliffs, 1985, p. 302.

[34] Prado, C. G. *The Last Choice*. Greenwood Press, Westport, Connecticut, 1998. p. xiv.

[35] Fletcher Joseph. "Sanctity of Life versus Quality of Life." In: Baird Robert M. and Rosenbaum Stuart E. (ed.), *Euthanasia*. Prometheus Books, Buffalo, New York, 1989, p. 85.

Chapter 6

DONATING OR SELLING ORGANS

The idea of legalizing the buying and selling of human organs for transplantation is frequently raised in the contemporary era, mainly in the Western world which has the financial resources for buying organs. The proponents of this idea claim that legalization could reduce the problem of the scarcity of donors and donations, leading to the saving of human life or at least, significant improvement of the quality of life of those who receive the organs. Another claim made in favor of the legalization of the trade of human organs is that it may actually assist poverty-stricken individuals, or individuals in debt, to ameliorate their economic distress by allowing them to sell their organs.

Another less-common claim in favor of legalizing the selling of human organs is that this may reduce the terrible phenomenon where "living people have their organs stolen for their cash value, or even worse, are killed so that their organs can be 'harvested' and sold."[1] This claim is strengthened by "worrying reports (from several countries including Brazil, India, Israel, and the Philippines) of body parts being stolen both from cadavers and from living hospital patients. These parts are said to include whole eyes (not only the cornea), bone, skin, pituitary glands, and heart valves."[2] Even though this claim may present a strong utilitarian argument, I choose not to include this kind of illicit and malicious organ trade in my discussion. I do not want to cloud the legitimacy of such a complicated issue with such overtly evil and criminal behaviors, though I realize that they exist. A more official and formal reason is the assumption that "Both ethics and the law in the West recognize the importance of respecting self-determination

---

[1] Wilkinson Stephen. *Bodies for Sale.* Routledge, London, 2003, p. 101.
[2] Wilkinson brings a list of sources for the above suspicions on p. 230, endnote 3.

or autonomy when it comes to people capable of making their own choices,"[3] as Timothy Murphy reminds us. However, I use the informed consent doctrine as only a formal argument in support of my view; my true reason for ignoring this claim is its malicious background, which makes it difficult to provide sincere arguments for or against it. When the basic assumption is that "Every human being of adult years and sound mind has a right to determine what shall be done with his own body,"[4] it is clear that harvesting organs without the donor's consent is illegitimate and evil and I see no need to dwell on that. So, I content myself with examining the situation where both seller and buyer traffic in human organs on a purely voluntary basis.

Legalization may seem to be a win–win situation as the interests of both sides can be promoted—those who need human organs for transplantation, and donors who need financial remuneration—leading to an improved quality of life of both sides. However, in order to reduce the danger of abuse of such legalization by speculators, profiteers or criminals, the demand to legalize the traffic in human organs usually includes the demand that this must be operated and managed by the State or other official health authority. This health authority would have to create an orderly, ethical, and acceptable framework for regulating the acquisition and transplantation of human organs for therapeutic purposes, and prevent coercion, exploitation, and abuse of such arrangements.

However, we must make an essential distinction between the selling of two kinds of organs: organs from a live, presumably healthy person and organs from a dying individual. Of course, there are limits to the type of organs a healthy individual can donate without compromising that person's own health, while many of the organs needed for transplantation can only be harvested from a dying person (such as the heart, liver, pancreas, and both lungs). We also stipulate that the buying and selling of organs would not be left to the open market, but confined to some kind of official authoritative health body. This health authority would distribute the organs through impartial considerations without charging the parties involved, which means that the government would cover the costs involved and not the persons who receive the organs (nor their health insurance). These stipulations must be taken into account when we examine the idea of allowing the trade of human organs.

Another stipulation for harvesting human organs from healthy people is that it must be confined to adults who explicitly express their informed and autonomous consent. We must never accept the consent of a person's legal custodian or relatives, in order to reduce the risk of exploitation of minors and mentally retarded

---

[3] Murphy Timothy F. *Case Studies in Biomedical Research Ethics*. The MIT press, Cambridge Massachusetts, 2004, p. 53.
[4] Dickenson Donna. *Risk and Luck in Medical Ethics*. Polity Press, Cambridge, 2003, p. 66. Dickerson cites there from Schloendorff v. Society of NY Hospitals (1914), but this is the general attitude of the US law to the doctrine of informed consent.

people who are more easily coerced or manipulated. In addition, we should confine the buying and distribution of organs to governmental authorities and ban the commerce in organs by private agents or bodies, since the latter would cater to wealthy people who can afford to buy the organs and while discriminating against the poor. Governmental authorities would, presumably, reduce the effects of economic inequality in the use of human organs for transplantation.

However, this is only a partial solution to the problem of the immense economic inequality between people, because it is obvious that in those countries where selling kidneys is legal "mostly the sellers are poor and healthy, whereas the purchasers are rich and unhealthy."[5] The vast majority of healthy people who consider selling their organs do so only because they are in dire financial straits or have huge debts. They are willing to harm their basic right to bodily integrity out of purely financial considerations. Thus the legalized traffic in organs has a heavy and undesirable social price: the bodily integrity of poor people is measured and evaluated differently from that of the wealthy. Any benefit to be gained from allowing the trade of human organs would be overshadowed by the resulting inequality between people as regarding their rights to bodily integrity.

We cannot easily pinpoint the guilty party in the trading of human organs: the donor, the purchaser, the broker making the deal or the physician involved. However, it is clear that the ethically unacceptable factor in this traffic is the exploitation that "is implicit in a market place situation."[6] Behind this assumption of exploitation lays the sad truth that "organ vendors' donations are not truly voluntary."[7] Even when the sellers are not physically coerced or intimidated, their reasons for selling their organs are not rooted in altruism but generally in economic factors such as poverty. Thus, even if their decision is fully autonomous, we cannot consider it to be fully voluntary. It can still be viewed as exploitation simply because "one person takes advantage of the misfortune of another for his or her own benefit."[8] The fact that "markets are somehow exploitative is perhaps the commonest way of arguing against them."[9] When we accuse someone of exploiting another person, this includes not only situations where the exploiter makes wrongful use of the exploitee, but also situations where "the exploitee or some part or aspect or behavior of the exploitee, makes a contribution to the achievement of the exploiter's goals."[10] This itself is malicious, and it also raises the fear that the large sums raised in these whole transactions may benefit mainly the brokers or middlemen, while the donors receive only a small percentage of the fees involved.

---

[5] Campbell Alistair, Charlesworth Max, Gillett Grant, and Jones Gareth. *Medical Ethics*, Oxford University Press, Greenlane, New Zealand, 1997, p. 57.
[6] ibid.
[7] Wilkinson Stephen. ibid. p. 105.
[8] ibid. p. 130. Here Wilkinson Kahn's position.
[9] ibid. p. 9.
[10] ibid. p. 26.

Moreover, the economic inequalities that may enable such exploitation are not only between different individuals from different social classes, but also between different states and continents. One of the objections to the trading of human organs refers to the fear of traffic in organs from Third World to wealthier countries. Lewis Petrinovich cites a report from February 25, 1994, published in the *Boston Globe*, in which the United Nations Commission on Human Rights found that "more people in India sell kidneys to strangers than in any other country," a finding that caused the Indian Parliament to pass "a law banning the sale of human organs and prohibiting the removal of organs from a living donor unless they are intended for a close relative."[11] The fear of exploitation here is not only between individuals, but also between continents. This raises another dimension of exploitation and injustice between the First and Third worlds: not only in lower wages and cheap labor, but even regarding bodily integrity.

However, while exploitation is a common hazard in any form of brokerage for profit (and to which all decent people should object), this particular form of organ exploitation involves bodily integrity and is, thus, simply intolerable.

There are two additional dangers involved with the trading of human organs. The first is the risk of objectification of the human body; Stephen Wilkinson brings Martha Nussbaum's analysis of the term, which says that "in all cases of objectification, what is at issue is ... treating one thing as another: One is treating *as an object* what is really not an object, what is, in fact, a human being."[12] There are several ways of objectifying persons, and Wilkinson, following Nussbaum, lists them as follows:

1. *Instrumentality*—the person is treated as merely a tool for, or means for, achieving the objectifier's goal.
2. *Denial of autonomy*—the person's autonomy is not recognized or not respected.
3. *Inertness*—the person is treated as lacking in agency ('and perhaps also in activity').
4. *Fungibility*—the person is treated as interchangeable with other similar persons (or even with relevantly similar non-persons).
5. *Violability*—the person's right to bodily integrity is not recognized or not respected.
6. *Ownership*—the person is treated as property.
7. *Denial of subjectivity*—the experiences of the person are not recognized or not regarded as significant."[13]

---

[11] Petronovich Lewis. *Living and Dying Well*. The MIT Press, Cambridge Massachusetts, 1998, p. 100.
[12] Wilkinson Stephen. ibid. p. 27. Wilkinson quotes from: Nussbaum M. "Objectification." *Philosophy and Public Affairs*, 1995, Vol. 24, p. 256.
[13] ibid. p. 28. Wilkinson quotes from: Nussbaum M. "Objectification." *Philosophy and Public Affairs*, 1995, Vol. 24, p. 257.

Donating or selling organs

One or more of these can objectify a person, and in the case of selling organs from a living person, there is a fear that more than one of the above exists. So, if we want to avoid exploitation, we should not allow the selling of organs by healthy people.

I now turn to the other option: the trade in organs from dying people who which will soon die anyway (for example after a horrific car accident), or even from a cadaver, when the selling of the unharmed organs can save other people's lives. For the sake of simplifying the discussion here I presuppose that while the organs may be sold, only official authorities can buy the organs and distribute them freely according to criteria of need. I also assume that the injured persons explicitly expressed their consent in the past that their organs may be sold if they reach a situation where they will not be able to survive; this is to satisfy my previous strict demand for informed consent. Even in cases such as these, it is more likely that poor people will sell their organs (due to their hope that their families will enjoy improved quality of life after their deaths, assuming that death is unavoidable in any case). Unlike the previous discussion about the harvest of organs from living persons, the current scenario poses no harm to the dying donors' rights to bodily integrity since death is imminent in any case whether or not organs are harvested from their bodies.

It would seem, then, that both sides would benefit from this trade: the patient who needs the organ for transplantation, and the family of the dying person, whose death is imminent in any case. The only institution to assume the cost of this trade would be the State, as part of its obligation to save its residents' lives, and we assume that donors have explicitly expressed their informed consent in advance.

However, I still voice my objections to the traffic of human organs from value, not utility, considerations. A product is usually ascribed at least two values: its use value, and its exchange value (which is its market or trade value). For example, the use value of a pair of shoes is protecting the feet, and it exchange value in exchange trade can be, for example, one coat (whose use value is thermal isolation, or protecting the body). In free trade its value may be, for example, $100 US. But with regards to the human body, we always try to avoid its trade value in order to distinguish between a living human being and a mere commodity or product. When we ascribe merchandize value to a human being, we may blur the distinction between a human organ and an animal's organ (or other commodity or product). This will harm the unique status of human beings as having intrinsic or sacred values. In Wilkinson's terms this phenomenon can be considered as commodification, which is "treating persons as if they were commodities."[14] That is to treat as fungible something ought not to be viewed as such. When one commodifies a person this person "breaches the second Kantian principle, since treating persons as fungible fails properly to respect their dignity; it is to regard

[14] ibid. p. 46.

them as having mere 'price'."[15] Thus, in order to preserve the intrinsic or sacred value of the human being, we must safeguard the current situation where we allow only the donation of organs but not their selling.

One can argue that commodification of a person or of a person's body is not relevant in this case, as we only refer to a dying person whose organs for transplantation can save other human life. I now raise two additional objections to the selling of cadavers' organs. I believe that Paul Ramsey was correct to assume that unless the door to monetary remuneration is barred, the right of the dead person's relatives will turn to be "ownership of the body for commercial purposes. No longer will this be a 'quasi' property right in the bodies of the dead, but full property rights for commercial purposes—at least in the body's 17 potentially usable organs and tissues."[16] The survivors may have gained lucrative asset in the body, and this will turn the generous act of saving life into mere trade or commerce. The noble act of bestowing life on other people may be subordinated to the market rules of supply and demand, and mainly to the principle of maximizing profit. This reduces the moral and social values of donating organs.

My second objection is that of social solidarity. Ramsey argues that "a society would be more civilized in which men are joined together routinely in making cadaver organs available to prolong the lives of others, than one in which this is done ostensibly by consent to 'gift' but actually for the monetary gain of the 'donor'."[17] Any social interaction for saving life through cadaver organs that is based on financial profit, would turn into a profiteering market of making money from dying members of the society. We should not slight what Ramsey calls "the potentially dehumanizing abuses of a market in human flesh."[18]

Yet, there is no doubt that people should consider the altruistic stipulation of donating their organs in case of a disaster, in order to save another's life. However, economic considerations should play no part in this decision and the decision to donate part of one's body, should be based on solidarity or sympathy or even altruism, but not on financial benefit of the living relatives. I can only agree with Paul Ramsey's quotes from Dr. D. Moore, who said that tissue transplantation may in point of fact give "an entirely new meaning to human generosity as living persons or families of those recently dead make free donation of tissue for the assistance of others."[19] Asking money for the honor of saving or improving the quality of someone else's life will annul the generosity involved in this act, and replace it with cupidity. This itself is regrettable and should be avoided. Here I

---

[15] ibid.
[16] Ramsey Paul. *The Patient as Person.* Yale University Press, New Haven, 1970, p. 213–214.
[17] ibid. p. 215.
[18] ibid.
[19] ibid. The quotation is from: Moore Francis D. "Ethics in the New Medicine." *The Nation*, 2000 (April 5, 1965): 365, and *Canadian Hospital* 42: 90.

do not even enter the argument that says that "the sale of organs might drive out donations and undermine the altruistic giving that is presently encouraged."[20] I would not use this argument even if it was based on solid evidence because I am not bothered by a possible decrease in the number of donations as much as by the essence of trade in human organs, which reduces the intrinsic value of human life. We should object to the very idea of selling organs per se, and not only because of mere utilitarian considerations.

However, there are still changes that should be made in the procedure (as accepted in the USA) of collecting organs from cadavers. Lewis Petrinovich mentions a regrettable norm he noticed in Kamm's book, where:

> "an original donor's decision to give an organ can be overridden after the death by the family's decision not to give, a practice that is contrary to legal proceedings regarding the distribution of the decedent's other goods... If the decedent has left no instructions, then the family is free to decide on the disposition of the remains but the family cannot override the decedent's decision *not to* donate organs. Medical stuff is reluctant to honor an individual's decision to donate organs without agreement by kin."[21]

My suggestion is to change this norm to be more humane. On the one hand I reject Harris' suggestion "that the consent of neither the deceased nor the survivors is required."[22] However, I think that family members should not be allowed to thwart the expressed desires of deceased persons who had stipulated their consent to donate organs after their deaths; such impediment by the family violates the donors' explicit wills and desires. Families should be consulted only in cases where the deceased have left no previous instructions on this issue; in those cases, the organs should not be considered "the property of the state upon death," as Dowie suggests.[23] We should respect the extreme emotionality attached to the occurrence of death and accept the family's preference when the decedent left no instructions. And when the decedent left explicit instructions *not* to use his organs after death, we must respect those instructions as well and not attempt to override them by obtaining the family's permission to harvest the organs. All human beings have ultimate authority over their bodies, even after death. However, when we do have the decedent's permission to donate the organs after death we should ignore the family's objections and use the organs for transplantation. This will show respect to the decedent's will.

---

[20] Petronovich Lewis. ibid.
[21] ibid. p. 91. the details are from: Kamm, F. M. *Morality, Mortality: Vol. I.* Oxford University Press, Oxford, 1993.
[22] The quotation is from ibid. p. 92, where he brings Harris view in: Harris J. *Wonderwoman and Superman,* Oxford University Press, Oxford. 1992.
[23] ibid. where he brings Dowie's view in: Dowie M. *We Have a Donor*. St. Martin's Press, New York, 1988.

Another norm (again, in the USA) which I believe should be changed is the one in which "A pregnant woman who chooses to have a legal abortion, however, is not allowed to donate the tissue of her aborted fetus to be transplanted to prevent a person's death."[24] This norm appears outrageous and arbitrary considering the scarcity of organs for transplant, even from embryos, thus in effect letting people die who might otherwise be saved. A woman who wants to ameliorate her grief involved in the abortion by giving life to another person through the donation of tissue from her aborted fetus, should have the opportunity to do so. Preventing her from doing so is contemptible.

I would like to sum up this discussion by citing the *Statement on Human Organ and Tissue Donation and Transplantation* made by the World Health Association (WHA), which was adopted by the 52nd WHA general assembly, Edinburgh, Scotland, October 2000: (paragraph 34).

> "Payment for organs and tissues for donation and transplantation should be prohibited. A financial incentive compromises the voluntariness of the choice and the altruistic basis for organ and tissue donation. Furthermore, access to needed medical treatment based on ability to pay is inconsistent with the principles of justice. Organs suspected to have been obtained through commercial transaction should not be accepted for transplantation. In addition, the advertisement of organs should be prohibited. However, reasonable reimbursement of expenses such as those incurred in procurement, transport, processing, preservation, and implantation is permissible."[25]

The following paragraphs examine a specific case of organ donation: donation of ovum.

CASE 1

I want to deal with two different examples related to the removal of ovums from women's bodies for purposes of fertilization. The first example is an advertisement made by a rich couple in a student's magazine of a prestigious university in the USA. This couple offered to pay $100,000 for ovums from a white sportswoman under the age of 30 who has high grades in the university. The second example is an Israeli woman who complained that many of her ovums where taken out of her body during her fertilization treatment without her consent, and then were transplanted in other women's bodies.

These examples show that the procedure of removing ovums from women's bodies during their fertilization treatments (as well as the use of these ovums) should be legally regulated. Currently, many countries allow ovums to be taken

---

[24] Petronovich Lewis. ibid. p. 103.
[25] Quotation is taken from: http://www.wma.net/e/policy/wma.htm. Also mentioned in: Wilkinson Stephen. *Bodies for Sale*. Routledge, London, 2003, p. 105.

from women during their fertilization treatment and some of these ovums are fertilized and transplanted back in the woman's body, while the rest of unfertilized ovums may be donated, with the woman's consent, to other women. Otherwise, women who are not undergoing fertilization treatment are flatly prohibited from donating ovums. This situation creates an acute scarcity in women's ovums and opens the door to speculation and profiteering of ovums, and also makes the fertilization treatment very expensive.

It is obvious that such a situation should be changed immediately. Women who do not undergo fertilization treatment but want to donate ovums anyway (voluntarily, and without pay) should be allowed to do so; barring them is an unjustified violation of their freedom and their rights over their bodies, and causes additional misery to other women who need the ovums. We should allow donations of ovums just as we allow donation of kidneys or corneas. On the other hand, when women undergo fertilization treatments in private medical institutes and willingly donate their excess ovums, these institutes extract payment from the women for treatment, and then also from the women who later receive these donated ovums. None of this money goes to the donating women; someone else profits from their generosity. Thirdly, the women who agree to donate ovums face a heightened threat to their health since they require an intensified round of hormonal treatments to produce additional ovums for other women, and often they are not informed of this health risk involved in ovum donation.

It should be absolutely prohibited to use a woman's ovums without her consent, and should be regarded just as reprehensible as harvesting any other organ without the consent of the donor. It should be unequivocal that "informed consent is important for ethically acceptable removal, storage and use of tissues,"[26] and this definitely includes the removal and use of ovums. The law should set compulsory punishments for doctors who are involved in such acts, and impose high compensation for women whose ovums are used for transplantation in other woman's bodies without their explicit consent.

The more complicated issue about the proposed idea to allow the donation of ovums is whether to also allow the selling of ovums. The proponents of commerce in ovums argue that this is the only way to reduce the scarcity of ovums for fertilization. They believe that payment might encourage those women who are not moved by altruistic motives. The supporters of this idea use the analogy of sperm donations, which are usually paid for, and many young people such as students do it and provide the necessary amount of sperm for those who need it. They argue that even though we euphemistically refer to the vendors as donors,[27]

---

[26] O'neill Onora. *Autonomy and Trust in Bioethics.* Cambridge University Press, Cambridge, 2002, p. 147.
[27] See ibid.

we should remember that this transaction is merely the selling and purchasing of organs under the norms of free market.

Although this analogy of the donation of sperm and ovum for money sounds reasonable, I believe that it is mistaken and misleading. First, I hold that there is no solid argument that supports or legitimizes payments for sperm donations. Although this practice has become accepted and taken for granted as morally justified, the sources of such legitimacy are not convincing. If we hold that trafficking in organs turns human beings into merchandize with exchange values, thus harming the moral status of life as having intrinsic value, then even the payment for human sperm is problematic. Nevertheless, this is not the main reason for rejecting the analogy between sperm donation and ovum donation. The more significant reason is that sperm donation does not harm the donor or put him in any physical or health danger, neither at the time of the donation nor in the future. The health risks involved to the woman in ovum donation, on the other hand, are much more serious—particularly under the current regulations.

In the usual discussion about organ donations by living donors, we classify donation into two categories "based on the likelihood and degree of risk and harm: donations which entail minimal to a minor increase over minimal risk, and donations which entail more than a minor increase over minimal risk."[28] This classification depends on different factors, such as the procedural risks that the harvest entails, the reversibility of the donation, and the long-term morbidity.[29] From this point of view, sperm donation does not include any risk during the harvest, the reversibility of the donation lasts several hours, and there is no risk of long-term sequelae. Thus, we can say that no physical harm is caused to the donor.

The harvesting of ovums presents a dramatically different story for the woman donor. This procedure involves risky hormonal treatment, requires general anesthesia, and increases the chances that the woman undergoing the procedure may develop ovarian cancer and other diseases in the future. These risks require a different stance towards the issue of ovum donation, and raise the fear that women, especially poor women, might risk their health to sell their ovums out of economic hardship and not altruism. Therefore we should ban the selling of ovums and only allow their donation, just as I feel we should ban the sale of other body organs. The scarcity of ovums should be solved by persuasion and not by the commercialization of this problem, and we should maintain our policy that "human tissues ought not to be bought or sold."[30]

Ruth Chadwick suggests a Kantian argument opposing the sale of bodily parts, which is based on duties towards our own bodies. These duties are "grounded in

---

[28] Friedman Ross Lainie. *Children, Families, and Health Care Decision-Making*. Clarendon Press, Oxford, 1998, p. 112.
[29] See ibid. p. 113.
[30] ibid. p. 148.

the duty to promote the flourishing of human beings, including ourselves."[31] I will present her interpretation of Kant's *Lectures on Ethics*, and show that the deontological criticism against selling organs is solid and convincing. Chadwick quotes Kant and then says that "a human being is not entitled to sell his organs for money, even if he were offered 10,000 thalers for a single finger."[32] Kant insists that "we can dispose of things which have no freedom but not of being which has free will."[33] This is an obvious Kantian position which bans the selling of human beings since all human beings have free will, and as such are not to be sold. However, Chadwick elaborates this claim to ban even the selling of part of a person's body. She goes further than Kant and argues that "a man who sells himself makes of himself a thing and, as he has jettisoned his person, it is open to anyone to deal with him as he pleased."[34] This idea is similar to the dangers of objectification, discussed earlier in this chapter. When one allows the selling of oneself, one makes oneself a thing. Once you place human beings in a position where they can be bought and sold, or their bodies are merchandized, others could treat these people as objects. Thus Chadwick's interpretation claims that according to Kant, a human being who has free will should not be bought or sold, and human beings may not offer any body parts or organs for sale, as this turns them into mere objects for other people's pleasure. Chadwick quotes Kant's decisive position, saying:

> "man cannot dispose over himself because he is not a thing; he is not his own property. To say that he is would be self-contradictory; for in so far as he is a person he is a Subject on whom the ownership of things can be vested, and if he were his own property, he would be a thing over which he could have ownership. But a person cannot be a property and so cannot be a thing which can be owned, for it is impossible to be a person and a thing, the proprietor and the property. Accordingly a man is not at his own disposal. He is not entitled to sell a limb, not even one of his own teeth."[35]

Chadwick's interpretation of Kant's view is that a person's bodily integrity is a kind of a person's duty to him/her self. However, this conflicts with our desire to encourage people to donate their organs, though Kant's view would probably not rule out organ donation after death. I still feel that we can reconcile our desire to encourage people to donate their organs, even while alive, with Chadwick's interpretation of Kant's view. When the motivation for donating organs is not to

---

[31] Chadwick Ruth F. "The Market for Bodily Parts: Kant and Duties to Oneself." In: Almond Brenda and Hill Donald. (eds.), *Applied Philosophy: Morals and Metaphysics in Contemporary Debate*, Routledge, 1991, p. 288.
[32] ibid. p. 291. The quotation is from: Kant I. *Lectures on Ethics*. (translated by Louis Infield), Harper & Row, New York, p. 124.
[33] ibid.
[34] ibid.
[35] ibid. p. 165.

make a profit but out of reasons of altruism and mercy, there is no merchandizing or degrading of our body or our dignity. Any loss of our bodily integrity is compensated by the saving of another life or improving another person's quality of life.

There is a big difference in societal values between the free donations of organs vis-à-vis the selling of organs for profit. Chadwick mentions Richard Titmuss, who argued that people should be encouraged to donate blood in Britain and not receive financial remuneration, and said "once the market ethic pervades that relationship it will have dire consequences for society's values as a whole."[36] When Chadwick reassesses Titmuss' claim after 15 years, she says about Titmuss' prediction, "At the time he wrote he thought there was a correlation between the values of the voluntary blood donor system in Britain and those of the National Health Service. With the undermining of the latter in the late 1980s, and the consequent dangers of creeping commercialism, it is even more important to uphold donation as opposed to selling."[37] We can see that the standards and norms of the market, which are basically exploitive, become monstrous when the merchandize is a human organ. Thus we should maintain these norms even with regard to sperm and ovum donations. I believe that the UK's *Human Fertilization and Embryology Authority* (FHEA) arrived at the correct decision when it stressed its commitment to the altruistic donation of sperm or eggs to create new life; that such donation should be a "gift, freely and voluntarily given."[38] The fact that we have arrived at a situation where we need to buy and sell sperm and ovums shows that something has gone terribly awry with human solidarity and altruism, and this should be changed. Sperm and ovum donations may be a good place to start re-educating the public regarding the importance of donations, thus rehabilitating social ethics of human solidarity and altruism, since both donations do not require too much sacrifice from the donors. If we succeed with these, we may be in a better position regarding societal encouragement of donation of other organs and not for money. This alone can be another reason to insist on the donation of sperm and ovum, and not their selling.

---

[36] ibid. p. 296. The quotation is from: Titmuss Richard G. *The Gift Relationship: From Human Blood to Social Policy*, Allan & Unwin, London, 1970.
[37] ibid.
[38] http://www.hfea.gov.uk/HFEAPublications/AnnualReport/1999%20Section%201%20HFEA%20Annual%20Report.pdf. This is from the eighth annual report 1999 of the association, p. 28. See also Wilkinson Stephen. *Bodies for Sale*, Routledge, London, 2003, p. 108.

Chapter 7

GENETIC ENGINEERING AND REPRODUCTION

The field of genetic engineering encompasses the seemingly insurmountable gap between technological progress and the inability of the ethics dictionary to respond to issues that arise in the wake of these developments in technology. The immense implications of genetic engineering on our moral thinking is so intense that according to Jurgen Habermas, it "changes the overall structure of our moral experience."[1] Habermas cites Ronald Dworkin's explanation for this change, as "the change of perspective which genetic engineering has brought about for conditions of moral judgment and actions that we had previously considered unalterable."[2] The quotation he brings from Ronald Dworkin stresses the dramatic implications that genetic engineering has for our traditional moral thinking.

> "We distinguish between what nature, including evolution, has created...and what we, with help of these genes do in this world. In any case, this distinction results in a line being drawn between what we are and the way we deal, on our own account, with this heritage. This decisive line between chance and choice is the background of our morality...We are afraid of the prospect of human beings designing other human beings, because this option implies shifting the line between chance and choice which is the basis of our value system."[3]

---

[1] Habermas Jurgen. *The Future of Human Nature.* Polity press, Cambridge, UK, 2003, p. 28.
[2] ibid.
[3] ibid. This idea is also presented. In: Dworkin R. "Playing God: Genes, Clones and Luch." In: Dworkin R. *Sovereign Virtue: The Theory and Practice of Equality*, Harvard University Press, Cambridge, 2000, pp. 427–452.

The most far-reaching debates in this area have to do with human reproductive cloning and stem-cell research. The conflicts between scientific demands and political or philosophical misgivings and qualms, sometimes make the issue intractable. We try to offer guidelines for therapeutic stem-cell research while, at the same time, banning human reproductive cloning, thereby attempting to address both sides of the dilemma consistently. A related issue, also discussed in this chapter, refers to some advantageous consequences of genetic engineering and stem-cell research in curing or preventing diseases. It is clear that future scientific/technological developments may, nevertheless, obligate us to rethink our own judgments—in this, as in all other issues in this section.

Before turning to the ethical discussion about possible benefits and dangers of stem-cell research, I want to briefly survey the three areas of research where stem-cells can be used. Stem-cells are found in certain body tissues and organs from the earliest stages of development [embryonic stem-cells (ES cells)] to adulthood (adult stem-cells). What makes them unique is that they can reproduce themselves infinitely and generate more specialized cell types such as muscle, nerve, or bone cells. ES cells are responsible for generating all tissues necessary for a baby's growing body, and adult stem-cells act as a repair system, continually replenishing ageing tissues with healthy cells. This ability to create new body tissues is what makes stem-cells such an exciting prospect for medical science. They offer the possibility of generating new cells to replace diseased and damaged body tissues in illnesses such as Parkinson's, diabetes, cancer, and Alzheimer's.[4]

There are basically three types of ES cells. The first type is the zygote itself and the first eight cells created by its three cell divisions; these are called totipotent stem-cells that generate the 216 different cell types that comprise the entire human body. Thus totipotent stem-cells are capable of developing into a complete human being. The second category are those cells created during the first 5 days of the growth of the original cells; these are called pluripotent ES cells which create a cluster of around 50 cells that form the tissues of the embryo. They can create most embryonic cell types but not all the tissues required for complete development. The third category are the partially differentiated cells called differentiated multipotent stem-cells; these persist in small numbers in some adult tissues and are capable of forming a limited number of specialized cell types, thus able to replace fully differentiated cells that are lost by depletion or damage. These cells can create bone-marrow, or replace injured limbs.

Stem-cell research uses mostly the ES cells because of their relatively large numbers and because they are much easier to manipulate than adult cells. This type of research aims to use stem-cell transplants to repair damaged and diseased body parts with healthy new cells. There are three potential methods of obtaining human pluripotent ES cells. The first method is by isolating them from surplus

---

[4] http://www.mrc.ac.uk/index/public-interest/public-topical_issues/public-stem_cells.htm

blastocysts created during IVF treatment (otherwise these blastocysts would be destroyed). The second method is extraction of cells destined to form eggs or sperm from aborted or miscarried fetuses. The third is the production of tailor-made ES cells from a patient's own differentiated adult cells using a cloning technique.[5]

One specific research field within general stem-cell research is called therapeutic cloning, also sometimes known as cell nuclear replacement (CNR). This research combines CNR with human stem-cell culture and stem-cell therapy. Its aim is to take healthy cells from a patient, reprogram a cell nucleus by CNR to grow pluripotent embryonic cell clones, and then induce them to differentiate into the stem-cell or mature cell types required for transplantation to treat disease. At this stage of the research, CNR is the only available means of reversing the differentiation of adult cells and restoring their embryonic potential, in order to generate perfect-match transplant tissue. The Donaldson Report, which was accepted by the British government and then by the parliament and the House of Lords, allowed the use of embryos, whether created by IVF or CNR (but not whole embryos beyond the 14-day stage) for researching treatment of childhood and adult disease and injury.[6]

However, while there seems to be opportunity for promising benefits from stem-cell research in general, and therapeutic cloning in particular, there is also opportunity for some terrifying consequences of cloning in the human domain. This fear was expressed by Jurgen Habermas in his outstanding book *The Future of Human Nature*. There he cites N. Agar's warning about the possible dangers involved in biotechnology, saying that "Science so often confounds the best predictions and we should not risk finding ourselves unprepared for the genetic engineer's equivalent of Hiroshima."[7]

The two examples I discuss in this chapter allude to both possible consequences of stem-cell research and its applications: the blessing as well as the curse.

CASE 1

In February 2005, Ian Wilmut of the Roslin Institute in Edinburgh (head of the team that created Dolly the sheep) and Professor Christopher Shaw (of the department of neurology at King's College London), were granted a license to clone human embryos for medical research of motor neurone disease (MND).[8]

---

[5] http://www.mrc.ac.uk/pdf_stem_cells.pdf
[6] http://www.mrc.ac.uk/pdf_therapeutic_cloning.pdf
[7] Habermas Jurgen. ibid. p. 118, note no. 5.
[8] Taken from Guardian Unlimited: http://www.guardian.co.uk/theissues/article/0,6512,1146550, 00.html

They were licensed by the British Human Fertilization and Embryology Authority, the same authority that earlier allowed scientists from the University of Newcastle to clone human embryos. Their research—which aims to find a cure for a host of incurable diseases, such as Alzheimer's and Parkinson's diseases and diabetes—nevertheless provoked fury from anti-abortion groups. The granting of these licenses by the Authority followed in the wake of the August 2000 recommendation of the Ethics Commission of the British Department of Health to permit the research of therapeutic cloning through nuclear-cell transfer. The next step was on December 13, 2000 when the British Parliament permitted the research of non-reproductive cloning (by a majority of 192 MPs) and the House of Lords approved it on January 22, 2001. In order to avoid misunderstanding and ensure that the legislation would not be interpreted to allow reproductive cloning, it was clearly stressed that stem-cells could only be retrieved from embryos up to 14 days old.

Stem-cell research and therapeutic cloning open a window of hope for those who suffer from neuro-degenerative diseases, and perhaps 1 day could even create live organs or limbs for transplantation. In addition to circumventing the problem of scarcity of donations, it would obviate the need for ensuring genetic harmony between the donor and the recipient since the required organ or limb could simply be reproduced through cloning. Another benefit which might be obtained by therapeutic cloning is using a patient's own bone-marrow to cure neuro-degenerative diseases such as Alzheimer's, Parkinson's, multiple sclerosis, and strokes, as well as many other diseases such as sickle-cell, anemia, cirrhosis, hepatitis, arthritis, osteoporosis, diabetes, and mitochondria. These possible benefits convinced the legislators in Britain to allow stem-cell research for the purpose of therapeutic cloning.

The debate regarding the legitimacy of therapeutic cloning research was probably clinched in favor of the research, by the decisions of the British legislators. However, debate still rages regarding the legitimacy of using therapeutic-cloning technology for human reproductive cloning. There are still two remaining questions in this domain. The more pressing question regards the legitimacy of human reproductive cloning; currently, most theorists, politicians, and scientists are adamant in opposing it. The second question, which is much more complicated but rests on the answer we give to the first question, is whether or not to allow the research which makes human reproductive cloning possible. That is: whether or not we should expand the above recommendation of the Ethics Commission of the British Department of Health to allow scientists to develop the research in the direction of human reproductive cloning and not only therapeutic cloning.

The main argument of those who insist that this research should be continued is that scientists must be granted extensive freedom in their research and should not be restricted in any way. They believe that scientific and research freedom is something which should be fully protected and not controlled by social or state authorities or regulations. The proponents of scientific freedom say that although they realize that scientific freedom is not absolute, we need to have very convincing reasons for restricting it and we need to prove concrete or substantial damage due to specific research in order to demand its termination. When such evidence is lacking, we cannot demand to stop science from moving forward in any field. Another reason to justify the restriction of research of any sort can be an acute violation of basic or fundamental values or principles upon which we establish our moral beliefs, norms or even daily life.

However, such "real and acute damage" is usually only discovered after the fact and often only from a long-term perspective. Since human reproductive cloning is so unique that we do not have any previous or even analogous experience, we cannot even envision its long-term effects. Thus, consequentialist considerations or arguments cannot be used to justify banning research. The only kinds of consequential arguments we can raise about such unfamiliar and unprecedented issues are speculative arguments, and these are not very convincing.

In any event, before we relate to the second question we must first answer the first: whether we should allow human reproductive cloning. In other words, if we think that we should allow human reproductive cloning then we must furnish the technological means to carry it out, including cloning, and would not restrict the research that makes this possible (according to the Kantian principles which states that *Ought Implies Can*). Thus we should turn first to the discussion about the legitimacy of human reproductive cloning.

There is no doubt that that the decisions taken by various different political bodies have expressed something like a consensus against reproductive cloning, at least on the official level. And this consensus, which appeared immediately after the cloning of Dolly the sheep, was not cracked during the 8 long years that have transpired since. Many countries like Spain, Germany, Australia, and Denmark have already banned cloning legally, while many others, like Switzerland and Canada, are engaged in preparing laws to ban it. This consensus can also be seen in the public responses to Dolly the sheep, as described by Richard Dawkins: "From President Clinton down, there was almost universal agreement that such a thing must never be allowed to happen to humans."[9]

This utterance referred, inter alia, to the recommendation of a Presidential Ethics Commission appointed by President Bill Clinton in 1997. This commission recommended "a moratorium on the use of federal funding in support of any attempt to create a child by somatic cell nuclear transfer." It also asked private institutions to comply voluntarily with the intent of this moratorium. The members of the committee believed that any attempt to create a child by somatic cell nuclear transfer would be "an irresponsible, unethical, and unprofessional act." Thus, they even recommended federal legislation to prohibit human cloning, with the

---

[9] Dawkins Richard. "What's Wrong with Cloning?" In: Nussbaum Martha C, and Sunstein Cass R. (eds.), *Clones and Clones*. W.W. Norton & Company, New York, 1998, p. 54.

possibility of reviewing the issue after 3 or 5 years.[10] Many other bodies made similar decisions such as the "Group of Eight" economic summit (of June, 1997 in Denver, Colorado) which voted to oppose human cloning.[11] In November 1997, UNESCO issued a declaration that prohibits the use of techniques that oppose human dignity, such as reproductive cloning. In January 1998, the Council of Europe issued a declaration that forbids any form of intervention aimed at cloning any person, living or dead.[12] In September 2000, the European Parliament passed a resolution calling for a complete ban on research on human cloning. The European Parliament did not even make any distinction between reproductive cloning and cloning for therapeutic purposes; in fact, the decision called for a complete ban on European-Union funding to institutions that are connected with research into human cloning (including therapeutic).

Such a comprehensive consensus is somewhat surprising in light of the differences of opinion that usually accompany most ethical disputes, and especially when we consider that the phenomenon of human cloning already exists in nature as in the case of identical twins. Identical twins are genetically identical and share all their biological characteristics from the very moment of fertilization. I will attempt to express the essential, innate human beliefs that lend people to strongly disallow any tampering with human reproductive cloning.

Before I discuss the reason that people intuitively tend to oppose the very idea of human cloning under most conditions, I bring the one condition under which some people do tend to be more lenient in favor of its use: to allow barren couples to have biological offspring of their own through human cloning.[13] This is basically a needs-based utilitarian argument in which utility may be invoked in the strict sense of increasing happiness and joy in the world (at least to the barren couple, and allegedly with no cost or loss to others). The above claim can also be examined as an issue of parental rights, thus belonging to the realm of discussions within right-based theories; or it can be examined as an issue of needs or of welfare, thus belonging to the realm of theories of justice. I, however, want to deal with the above biological-offspring question as an issue of values, and examine it from a deontological point of view.

The issue of allowing barren couples to create a human being in order to realize their parental rights brings us face-to-face with the following moral dilemma: whether any aim or goal, no matter how positive, can be a reason to create human

---

[10] "Recommendations of the National Bioethics Advisory Commission." In: Nussbaum Martha C, and Sunstein Cass R. (eds.), *Clones and Clones*, W.W. Norton & Company, New York, 1998, p. 292.

[11] Kolata Gina. *Clone,* William Morrow and Company, Inc., New York, 1998, p. 229.

[12] ibid. p. 32.

[13] Another argument can be raised by biotechnology companies, which can make profit from providing such a service to people who need it. However, this is a much weaker argument than that of barren couples, so if we bring arguments against cloning which override of the main argument in favor of it, we should not be bothered by the weaker one.

life. I believe that a positive answer to this question might violate one of the most fundamental moral imperatives: Kant's categorical imperative in its humanity version. This imperative demands we must always treat the humanity of others as well as of ourselves as ends, and never as a means. If we allow the creation of a human being in order to achieve any goal or aim, no matter how noble, there is the fear that we might consider them (the new human being) to be means and not ends. This may be the fear that induces people to wholesale banning of human reproductive cloning, despite the sorrow involved in preventing people to have their own biological offspring.

One can say that many people who decide to give birth to a new child do it because they want to fulfill their own desire or will. This still does not cause those parents to treat their children as only means. Parents who want to use cloning techniques can say that what they do is not dramatically different from parents who use the IVF technique in order to fulfill their desire for children. At this point, I want to explain the relevant difference between cloning and IVF. In the case of IVF or even standard fertilization treatments, potential parents have no control over the physical characteristics of their offspring just as all parents who conceive a child naturally have no control. However, producing a cloned child means that the genetic characteristics of the cloned baby will be identical to one of the parents. Thus the parents ascribe more significance to the material or physical characteristics of the future person, thus reducing the importance of the mental or spiritual dimension of the clone. This, in term, might change our usual hierarchy between the two aspects of a person's existence. Instead of giving priority to personality and cognitive characteristics over physical ones, we might subordinate the mental and emotional characteristics of a person to the physical ones, thus eroding the pre-eminence of man "and may get to the point where we might treat people merely as having a certain set of physical characteristics, and not as autonomous individuals. The whole concept of an autonomous and free person, which is based on a person's spiritual and mental abilities, may be eroded by the concentration on a person's physical characteristics. This, again, may lead us to treat people as means and not as ends."[14]

In this sense, I believe that such "instrumentalization of human nature changes the ethical self-understanding of the species in such a way that we may no longer see ourselves as ethically free and moral equal beings guided by norms and reasons."[15] And this instrumentalization of human nature may initiate "a change in the ethical self-understanding of the species—a self-understanding no longer consistent with the normative self-understanding of persons who live in the mode of self-determination and responsible action,"[16] as Habermas puts it.

---

[14] Ezra Ovadia. "Human Reproductive Cloning: Hope or Disaster." In: *International Journal of Politics and Ethics*, Vol. 3, No. 2, 2003, p. 215.
[15] Habermas Jurgen. ibid. p. 40–41.
[16] ibid. p. 42.

Another aspect that is sometimes neglected in the discussion of the legitimacy of human reproductive cloning is that the child might not be able to escape the life history of his chronologically out-of-phase "twin"—that is, the parent whose genetics are reproduced in the child. Such 'programmed' persons might encounter two possible consequences: first, they "might no longer regard themselves as the sole authors of their own life history, and second... they might no longer regard themselves as unconditionally equal-born persons in relation to previous generations."[17] Even if a programmed person will not experience discrimination in his/her social surrounding, he/she will suffer "a prenatally induced self-devaluation;... a harm to her own moral self-understanding. What is affected is a subjective qualification essential for assuming the status of a full member of a moral community."[18] Such people might suffer from psychological alienation, dilution or fracturing of their own identity, and this may be a sign that "an important boundary has become permeable—the deontological shell which assures the inviolability of the person, the uniqueness of the individual, and the irreplaceability of one's own subjectivity."[19]

Now that we have reached the conclusion that human reproductive cloning should never be permitted, we still have to decide whether we should allow the development of the technology to enable such cloning. Those who want to safeguard scientific and research freedom under all circumstances, can argue that even if we reject human cloning there is no reason why we should ban the research regarding it. That is, they maintain that we can still develop technology while banning its use. At this point, I want to raise objections that are similar to the objections raised with regard to the development of the hydrogen bomb (the "Super"). This deterministic hypothesis maintains that when technology exists to produce something, even if this "something" is monstrous or might have dreadful effects, there is no power on earth that can stop the production of the monster. No person or group of people on earth who create a monster will be able to control it, or avoid its dissemination. This hypothesis was first raised by those who objected to the nuclear armament race and demanded the termination of research that could lead to the uncontrollable production of destructive weapons. They claimed that such weapons could threaten the continuation of the existence of mankind and even lead to omnicide. Yet the bomb was finally produced; people then understood that technological imperative is so powerful that the development of any technology is an irreversible stage, leading inevitably to the use of that technology—in this case, the production of the hydrogen bomb. From this analogy we derive that if we object to an end-product of a technology, or in our case human reproduction cloning, then ipso facto we must prevent the development of the technology which enables it. Thus we should ban

---

[17] ibid. p. 79.
[18] ibid. p. 81.
[19] ibid. p. 82.

Genetic engineering and reproduction   89

the development of the research which might make human reproductive cloning available.

A similar idea can be found in Habermas' quotation from W. van den Daele "That which science made technologically manipulable reacquires, from a normative perspective, its character as something we may not control."[20] The main fear in this domain is from the instrumentalization of human life, about which the President of the Federal Republic of Germany, Johannes Rau said in May 18, 2001: "Once you start to instumentalize human life...you embark on a course where there is no stopping point."[21] Habermas himself raises the fear that technology might have its own dynamics, when he says: "part of a moral reflection on legal policy, reference to the normative force of established facts will only confirm a skeptical public's fear that science, technology and economics may create, by their systemic dynamics, *faits accomplis* which can outstrip any normative framework."[22]

Occasionally, announcements are made from time to time about so-called success of human reproductive cloning. The most famous one was made by an Italian doctor, Severino Antinori, who announced in mid 2002 that he managed to fertilize a woman with a clone of an Arab adult person, and that she was supposed to deliver the baby in Belgrade Serbia in January 2003. This announcement is now believed to be a hoax. Some examples of successful experiments on animals include; a 6-year-old sheep that was successfully cloned from a mature cell in Scotland in 1996; a calf that was cloned from a mature cell in 1998; and a chimpanzee (which is about 99% genetically similar to the human body) that was cloned in 2000. All these examples, and especially the chimpanzee cloning, might mean that modern science already possesses the technology to clone a human being; some people even believe that a human being has already been cloned secretly. I personally do not believe that human reproductive cloning has been carried out, but since it is likely that science possesses the technology to do so, it is imperative upon us to do everything in our power to prevent the realization of this technology since human cloning is likely to have terrible consequences.

It was this fear of consequences that caused the Roslin Institute–pioneers in cloning of a mammal from an adult cell—to halt its development of special pigs from which it hoped to harvest organs for transplantation in human bodies. They were afraid that if such organs from pigs would be transplanted to humans, then diseases that previously targeted pigs might also attack people who accepted these transplanted organs, and then attack the human population at large. Although there was no actual proof that this might happen, it was enough of a scare to convince the scientists involved to halt the research.

---

[20] ibid. p. 25.
[21] Quotation is taken from ibid. p. 19.
[22] ibid. p. 18.

I believe that the dangers involved in human cloning far surpass the dangers involved in transplanting pig organs, and justify the ban on research for human reproductive cloning. In fact, the decision reached by the Roslin Institute follows what Gordon Graham calls "The Precautionary Principle" which says, in effect, that one must halt a course of action when the possible negative outcomes of this course of action (even far-fetched possible outcomes) are so catastrophic that "the merest prospect of their coming about as the result of something we do is enough to give us reason not to do it."[23] Although this principle is usually invoked in environmental debates (for issues such as nuclear winter and global warming), it can also be used for other issues of immensurable magnitude.

Another substantiation for justified fears of human reproductive cloning—and not only the result of an hysterical response to the unknown—lies in the life-history of Dolly, the first cloned mammal, from a mature cell. Dolly was cloned from a cell which was taken from a 6-year-old sheep, and when Dolly matured to the ages of about 5–6, it had syndromes and health problems that usually characterize much older sheep, such as arthritis. This may be a sign that although her chronological age was 6, her biological age was the accumulation of her age and that of the cloned sheep from which her first cell was taken. Dolly the sheep was mercy-killed in February 2003, after suffering from a progressive lung disease that usually attacks sheep twice her age. Scientists believe that "The key element was the telomeres, repetitive DNA on the ends of chromosomes that normally protect the ends. Every time a cell replicates its DNA, a little bit of the end of the telomere does not get replicated. Thus the telomeres shorten with age, which can cause certain age-related health problems. Dolly's telomeres were already shortened when she was born because they were from another sheep; this was the cause of her arthritis and may have played a role in her early death." This consequentialist argument raises the specter of cloned young children suffering from the arthritis and cardiac problems that plague their parents, and is another reason to ban cloning.

However, even if "for the foreseeable future . . . the cloning of a developed human being is out,"[24] we should still be worried of the possibility of its appearance from unexpected quarters and make efforts to prevent it. In my opinion, it is the primary responsibility of the scientific community to thwart both the research of human reproductive cloning and the cloning itself. Legislation cannot encompass the whole world seamlessly and there will always be cases of falling between the cracks—isolated areas with no legislation against cloning, within which this procedure can take place. An example of this happened when Australia originally passed a law banning in-vitro fertilization (IVF) within its territory. Australians who wanted to have IVF only had to travel to Singapore for a couple of days to have the, only the scientific community has the genuine ability to prevent or

---

[23] Gordon Graham. *Genes.* Routledge, London and New York, 2002, p. 128.
[24] ibid. p. 155.

Genetic engineering and reproduction 91

struggle against reproductive cloning, mainly by refusing to cooperate with rogue scientists who involve themselves in such research or its applications. Scientists can ostracize those scientists and scientific institutions that promote cloning, thus lowering the odds. However, the scientific community has a huge financial interest in continuing this research and, thus, I fear that it will adopt the position of the British Ministry of Health, which believes that the utility which can be gained from cloning overrides every ethical consideration. And when utilitarian considerations override ethical considerations, disasters are sure to follow.

CASE 2

My next case deals with questions that arise from *preimplantation genetic diagnosis* (PGD) procedure. This procedure permits genetic screening to be carried out on embryos at the eight-cell stage. Such a procedure is recommended to parents who wish to rule out the risk of transmitting hereditary disease. In this procedure, the embryo in the test tube is screened and if it carries the genetic disease, it will not be implanted in the mother's womb. This procedure spares the mother from undergoing an abortion at a later stage after prenatal diagnosis.[25]

The case I deal with here happened in Colorado, USA, in October 2000. Jack and Lisa Nash, both suffering from a genetic defect, had a 6-year-old daughter, Molly, who was born with Fanconi's anemia. This is a gene-related bone-marrow disease which threatened her life. They decided to take advantage of IVF technique and the possibility of genetic screening in order to make sure that their second child, Adam, would not carry the gene which causes that disease. However, they added an additional element: among the healthy embryos which were created during the fertilization treatment, they chose the one whose tissues matched his sister Molly, so that stem-cells recovered from Adam's umbilical cord might be used for transplant in Molly's body to increase her chances of survival. To some extent, we can consider this procedure to be *Negative Eugenics* since it "involves the elimination of defective genes, and hence, the prospective possessors of these genes from the population."[26] But we should remember that "negative eugenics is not an illustration of manipulation, since it does not modify individuals who will be born. However, it is a means of selecting healthy as opposed to unhealthy individuals."[27]

Adam was born in October 2000, and his birth immediately raised two ethical debates. One was whether it is moral to select an embryo for the benefit it can give, and the other concerned the number of embryos produced before a suitable

---

[25] See, Habermas Jurgen. ibid. p. 17.
[26] Campbell Alistair, Charlesworth Max, Gillet Grant, and Jones Gareth. *Medical Ethics*, Oxford University Press, Greenlane, New Zealand, p. 66.
[27] ibid.

match was found (over 100 embryos).[28] The first question refers to the specter of producing babies in order to be used as tissue reservoirs or storehouses for transplantation, something which clearly violates Kant's categorical imperative that human beings (including babies) must always be viewed as ends and never as only means. In order to examine such a threat, we have to make a few distinctions before we consider the above case as an attempt to use Adam, the newborn baby, as a tissue reservoir.

The real and substantial threat of using children as tissue reservoirs usually exists in cases where a significant and irreversible organ (such as a kidney) is harvested from a child (this was discussed in the previous chapter). Such a case harms, or at least threatens, the child's bodily integrity and places this child at risk; both during the harvesting procedure itself, and in the future (living with only one kidney for the rest of his/her life). The desire to improve someone else's quality of life or even save the other's life, cannot justify the harm and danger for the child from which the organ was harvested. Thus we rejected the use of children as organ donors (in the previous chapter) for several reasons, such as the child's fundamental right to bodily integrity. In my opinion, at least with regard to the tissue reservoir charge, the case of the Nash Family is categorically different. They definitely acted within their authority while using their son's organ for donation, particularly when this was an intrafamilial donation. Lainie Friedman Ross, who studied the issue of children as organ donors, maintains that "the role of a child as an organ donor for an intrafamilial transplant is justified in part because the intrafamilial donation not only uses the child as a means to help another family member, but it also serves to promote the family's well-being, on which the child's own well-being depends."[29] The fact that the donating child suffered no risk makes such a decision, according to Friedman Ross, justified even if the donation was not intrafamilial, but for an outsider.[30] In the coming paragraphs, I will try to highlight this distinction and examine the relevant moral consideration which justify the parents' decision, at least with regard to the criticism of "tissue reservoir."

Adam was born to parents who both carry a defective gene which probably caused Molly's disease. The parents claimed that they did not bring Adam into the world to cure his sister, but instead, wanted to have another child who would not carry the same defective gene and would not be ill. They were offered IVF treatments in which the created embryos could be tested in order to choose the healthy ones and transplant them in the mother's womb, thus ensuring that the newborn baby would not carry the defective gene.

This presentation of the story changes its moral judgment, due to several reasons. First of all, the classification and testing of potential embryos is justified

---

[28] http://www.wellcome.ac.uk/assets/wtd003616.pdf. p. 16.
[29] Friedman Ross Lainie. *Children, Families and Health Care Decision Making*. Clarendon Press, Oxford, 1998, p. 124.
[30] ibid. p. 125.

Genetic engineering and reproduction

in order to avoid inflicting additional misery on the newborn baby, the parents, and even the sick sister (who would conceivably also suffer from the additional burden of an ill sibling on the parents' shoulders). This process—of selecting a healthy embryo—has become a frequent technique among embryos that carry the risk of genetic defects such as cystic fibrosis. In our case, the procedure provides an additional bonus—a chance of curing the sick sister—and presents no moral problem. Even Jurgen Habermas, who is ordinarily very critical about the whole procedure of PGD, still believes that a consensus can be reached about it "for the goal of avoiding evils which are unquestionably extreme and likely to be rejected by all."[31] Thus even he would consider such a genetic disease to be an "evil" that should be "rejected by all."

The only remaining problem here relates to the demand for specific characteristics that are required to heal the sick sister. However, these "specific characteristics" are not morally objectionable ones such as hair color, height, or even intelligence quotient (IQ), but only the *lack* of a defective gene that would cause the baby much pain and grief. Genetic engineering operations of this sort have been widely accepted both by the scientific community and by the ethicists of stem-cell research. What is not accepted in the scientific community is the *adding* of a gene not carried by the parents, a procedure that can change the whole genetic composition of all future descendents of these parents. The great fear of the ability to screen embryos, as Herman Saatkamp reminds us, is not due to cases where parents want to screen out certain characteristics "but also to include traits in their children they consider positive. Such traits include not only socially privileged ones, such as body build, but perhaps traits of sociability, aggression, intelligence, sexual orientation, and more."[32] None of these traits apply to the Nash family preferences; their only desire was a healthy baby who could, possibly, also help his sister.

Another possible issue may be the number of embryos that are wasted until the suitable genetic composition is found; this issue hinges on the stage at which we ascribe "personhood" to a human organ. Since the selection of embryos is performed at a very early stage—soon after conception—only very religious approaches consider an embryo of this age as possessing personhood. I do not want to enter into a prolonged discussion here, but only to state my opinion that there is no personhood at this stage. Such embryos are not much more developed than sperm or ovum; they do not have any sensation and do not possess any cognitive apparatus; they are routinely discarded in all IVF treatments and, except for very religious groups, this is not considered a real moral problem. Since I, personally, do not consider in-vitro embryos as possessing personhood, their loss is not such a terrible moral issue that might change the morality of the parents' decision.

---

[31] Habermas Jurgen. ibid. p. 43.
[32] Saatkamp Herman J Jr. "Genetics and Pragmatism." In Mcgee Glen. (ed.), *Pragmatic Bioethics*, The MIT Press, Cambridge, Massachusetts, 2nd ed, 2003, p. 165.

There is still something to be said about the considerations which should be taken in the use of negative eugenics. Even if it "does not actually changing individuals, its potential for changing the genetic make-up of a community is considerable."[33] And since it is implicated in manipulation of the population, ethical considerations should be taken into account in decision-making about each use of negative eugenics. Alistair Campbell, Max Charlesworth, Grant Gillet, and Gareth Jones cite some of the most significant of such considerations. They mention "the severity of the genetic disorder and its effect on the possibility of meaningful life for the affected foetuses; the physical, emotional, and economic impact on family and society of the birth of a child with the genetic condition; the availability of adequate medical management and special educational facilities; the reliability of the diagnosis; and the increase in the load detrimental genes in the population as a result of the carriers of genetic diseases reproduction."[34] In any event, if there are possible benefits to be gained from negative eugenics and low risks (unlike positive eugenics where there are tremendous risks), we should try to attain these benefits, of course with the appropriate caution and constraints. We should make sure that we do not deviate or exceed the proper use of genetics, which according to Timothy Murphy is "to free human beings from the burden of disease."[35]

I want to clarify that although I stated that the scientific community bears the onus of responsibility to supervise and monitor genetic engineering research, this does not rule out the importance of governmental regulations. I do not agree with Nobel Price Laureate David Baltimore, a virology professor at MIT, who argued during a dispute over recombinant DNA, "The scientific community, being as open as it is, and as self critical, provides a better guarantee of safety than does any government regulation."[36] I tend more to the opposite approach, represented by Jonathan King (one of Baltimore's colleagues at MIT) who argued, "scientists are no more to be trusted to police themselves than the tobacco industry to determine the danger of cigarette smoking."[37] Thus we must retain the license to put legal constraints on scientists who deal with genetic engineering. This is because even if they have no malicious intentions, scientists "might well become so fascinated by the development of the procedures as to lose sight of this... laudable aim. For them science might become an end it itself."[38] They are likely to lose sight of the morality or even the dangers involved in their research, hence they should be regulated by an external agent. We should remember that "the advent of

---

[33] Campbell Alistair. ibid. p. 66.
[34] ibid.
[35] Murphy Timothy F. *Case Studies in Biomedical Research Ethics*. The MIT press, Cambridge, Massachusetts, 2004, p. 236.
[36] Kolata Gina. ibid. p. 112.
[37] ibid.
[38] Holland Stephen. *Bioethics*. Polity Press, Cambridge, UK, 2003, p. 200.

Genetic engineering and reproduction

genetic engineering... provides the potential for the design and redesign of living matter to human purpose. And living matter includes homo sapiens."[39] This inclusion requires the most decisive constraints available, such as Cynthia Cohen's recommendation of a public oversight body to monitor stem-cell research carried out throughout the country.[40]

However, this is not enough. The main responsibility still remains on society, and even on mankind as a whole. If we fail to regulate scientific progress, particularly in the domain of genetic engineering, we might experience the "horrifying prospect that a eugenic self-optimization of the species, carried out via the aggregated preferences of customers in the genetic supermarket (and via society's capacity for forming new habits), might change the moral status of future persons: "Life in a moral vacuum that not even the moral cynic would still recognize, would not be worth living."[41]

---

[39] Sinsheimer Robert L. "The Prospect of Human Genetic Engineering." In: Blank Robert H and Bonnicksen Andrea L. (eds.), *Medicine Unbound*, Columbia University Press, New York, 1994, p. 112.
[40] Cohen Cynthia B. "Expanding Oversight of hES Cell Research." In: Holland Susanne, Lebacqz Karen, and Zoloth Laurie. (eds.), *The Human Embryonic Stem-cell Debate*. The MIT Press, Cambridge, Massachusetts, 2001, p. 220.
[41] Habermas Jurgen. ibid. p. 95.

SECTION C. PARENTHOOD AND THE FAMILY

Modern—or is it post-modern?—life styles have created new forms of relationships between, within, among, and outside the traditional nuclear family (consisting exclusively of mother, father, and their own offspring). These new frameworks have given rise to moral rights and obligations that weaken the old patriarchal and absolutist structures of past traditional families. For example, grandparents have gained certain legal rights over their grandchildren in the United States, and single parenthood has become overwhelmingly accepted in many countries throughout the world. Another change in the traditional relations within families results from new technologies such as IVF (in vitro fertilization) sperm donation and DNA testing These technological innovations enable widows to bear children from their deceased husband' sperm, mothers to bear children from their deceased son's sperm, or simply allow infertile couples to bear children from sperm or egg donations of strangers.

Thus, this section is devoted to some of the issues that have arisen due to these changes, such as the rights of adults born as a result of donated insemination and the rights of grandparents and other extended family members vis-à-vis the parents. Our attempts to navigate between the conflicting claims of different parties of the extended family sometimes raise very complicated issues, and these issues are examined here.

CHAPTER 8: RIGHTS OF RELATIVES AND GENERATIONS

The presupposition of dealings in rights of family matters has always been that parents are the ultimate authority in anything having to do with (their own) children. However, the waning of traditional norms brings about challenges to this assumption. In this chapter, we address the rights of relatives beyond the nuclear family. The first issue discussed in this chapter are grandparents who insist on their rights in seeing, meeting, or maintaining contact and relationships with their grandchildren against the will and preferences of the parents (who are the grandparents' children). The issue of grandparents' rights has become accepted in US Courts as all 50 states have laws that acknowledge such rights, at least to some extent. My view regarding this issue acknowledges and respects these rights and includes them within the large framework of rights within the family. In my opinion, though grandparental rights are subordinate to parental rights, they should still exert considerable weight when there is a conflict. Of course, the decisive consideration in resolving the conflict of parental rights with rights of grandparents should remain the children's welfare.

The second issue of this chapter relates to the controversial case in year 2000 of the Cuban child rescued from drowning and brought to the US where his mother's relatives claimed custody over him against the claim of the Cuban father (the mother drowned in the attempt). Although the boy was rightly returned to his father in Cuba, this case exposed some ugly aspects of American society though ultimately, the superior status of parental rights over the rights of other extended family members was maintained.

The third issue discussed in this chapter is a new law in Britain which allows children who will be born in the future of sperm donations to reveal the identity of their biological fathers, who will no longer have the right to anonymity. Apparently this decision equates the rights of children who were born from sperm or egg donations to those of adopted children, who can see the adoption files when they turn 18 and become adults. (This will affect only future donors and children, and not those in the past whose anonymity will still be maintained.) In this chapter I discuss how such a law changes sperm and egg donations within the concept of biological parenthood.

CHAPTER 9: PROCREATION AFTER DEATH

A more complicated issue that relates to parenthood and family is the desire of parents or spouses who have lost loved ones to use the deceased's sperm in order to create a new generation of that same family. There are a number of subtle moral dilemmas: often, it is not technically possible to obtain informed consent of a dying man. Also, a child produced from the sperm of a deceased man, will be the biological grandchild of the man's parents who will raise him as their child, causing a skip-over or a confusion of generations. Finally, there is the dilemma of bringing a child into the world as a "monument" to a deceased family member.

Parenthood and the family 99

This chapter deals with two different requests, each with its own complexity. The first case is the request of a young widow to use the sperm of her deceased husband to have his child. Although the sperm was harvested from the husband before his tragic and unexpected death, he was not able to give informed consent as he was unconscious. However, his parents and the widow's parents all gave their consent and agreement to support the young widow in raising the man's biological child.

The second case is the request of parents to use the sperm of their dying son to create another child; in this case, the dying son was able to give his consent. The moral issue here is of a child born to much older parents who might not be able to take care of him appropriately over the long term, and might even need the new child to take care of them due to their advanced age.

CHAPTER 10: BABIES AS COMMODITIES

The dimensions of global trade between rich and poor countries—certainly a direct consequence of current globalization—has both transcended traditional state boundaries as well as transformed anything and everything into objects of trade and commerce. That human beings have been used as commodities is familiar in our history: witness slavery and prostitution. However, the extension of this regrettable phenomenon to babies is one of the most objectionable aspects of our new global form of life and thought. What started as a generous movement of international adoptions, with well-meaning motivation and intention, has deteriorated into a capitalistic profit-making venture in which babies are no more than a means of maximizing profits. In this chapter we endeavor to pose, and answer, queries about the ethical implications of the tragic move from adoption to baby commerce.

The specific case is a story about an adoption agency in San Diego that first offered twin girls for adoption to a couple from California and then took the babies back shortly afterwards to offer them to a couple from Britain, who evidently paid more than the first couple. The whole story was exposed after the babies arrived at the new home in Britain. This story ended after 4 years, when the twins were finally restored to the custody of their biological mother. But the entire case raises the specter that children are becoming just another form of merchandise for sale in the Internet, as "paid adoption" may cross the line into actual baby commerce. This chapter discusses the meaning of commodification of human beings in general, and of children in particular.

Chapter 8

RIGHTS OF RELATIVES AND GENERATIONS

The traditional extended family encompasses different generations (such as children, parents, grandparents), and different forms of kinship (such as children, parents, brothers, uncles, aunts, grandparents) and relations that are usually based on affection, love, support, and mutuality. However, relations even in the traditional family structure are not always typified by affection and love, but by exploitation, conflict, and hostility. In cases of a family unit gone awry, different members of the family may present conflicting claims for rights and the resolution of such conflicts may be left to the courts. Sometimes the disputes revolve around the right to keep contact with children, and frequently the children are the primary victims of the intrafamilial conflict. Sometimes the conflicts involve extended family members. This chapter examines cases of conflicts within the family, sometimes as a result of family crisis, and in other cases as the result of unusual circumstances that require creative and innovative solutions.

The traditional "nuclear" or "core" family is based on marriage between a male and a female. Today, although this is still the dominant form it is not the only one; in some countries, it is not even the most frequent or popular. David William Archard suggests a more updated concept that "a family is essentially a stable multigenerational association of adults and children serving the principal function of rearing its youthful members."[1] The familial relations are still characterized by intimacy, affective closeness, and unconditional love, but it might include more than two generations, more than one habitation, and the parents might comprise less than two genders. The rise of reproductive technology enables us

---

[1] Archard David William. *Children, Family and the State.* Ashgate, Aldershot, 2003, p. 69.

to distinguish between "natural" and "custodial" parenthood, and made it much easier for same-sex couples to have children.[2]

Other important trends in the family are: the decline of marriage and the rise in divorce rates; the rise in number of children born to cohabiting couples and the number of dependent children cared for by single parents and by step-parents. David William Archard reveals that "over one-third of all marriage are remarriages, one in five of all dependent children live in lone parent families, and one in four of all women aged between 18 and 49 are cohabiting."[3] So it appears that in most modern Western societies there are a variety of familial forms. However, the new family structures mean that familial relations are more complex and need to be re-examined, as we shall see in the third example of this chapter.

Since children are usually the most vulnerable links in the family chain, the courts are often called upon to regulate or settle conflicting claims of other family members regarding the children. Sometimes the courts rule against the parents' desires and sometime in favor of them. In any event, "it is a standard principle of child's welfare law and policy that the 'best interests' of a child should be promoted."[4] This principle (Best Interest Principle, or BIP) has become a cornerstone of legislation and in decisions of the courts, but in many cases the decision regarding a child's best interest is a matter of interpretation. The present chapter deals with some disputes and disagreements regarding the question of what is really the child's best interest, and presents examples to reveal the complexity of the issue.

CASE 1

This case deals with verdicts handed down by American courts allowing grandparents to visit their grandchildren over the parents' objection, such as *Blakely vs. Blakely* in Illinois. Today, "all 50 states currently have some type of 'grandparent visitation' statute through which grandparents and sometimes others (foster parents and stepparents, for example) can ask a court to grant them the legal right to maintain their relationships with loved children."[5] Even the US Supreme court decision in *Troxel vs. Granville*, which states that "parents have a fundamental right to make decisions about raising their children," does not hold that the permissive visitation statute is unconstitutional or that allowing a non-parent to petition for visitation rights would amount to "an assault on the integrity of the family unit"[6]—thus acknowledging, in effect, the visitation rights of grandparents.

---

[2] ibid, p. 71.
[3] ibid, p. 72.
[4] ibid, p. 38.
[5] http://www.nolo.com/article.cfm/ObjectID/1019223D-59A2-4A25-9B6DA99AD0406A0F/cat-ID/AC0903D2-C845-40E8-850E1DCEDDEA5778/118/246/236/ART/
[6] ibid.

Rights of relatives and generations

I discuss here the possible origins of visitation rights of grandparents as a moral right, while limiting my discussion to the classic family structure where the biological parents are also the custodial parents, and both parents and grandparents are alive and competent (not cases of divorced parents). But basically I want to challenge the commonly accepted idea that was reaffirmed by the US Supreme Court (Troxel vs. Granville) decision, which ascribes fit parents the ultimate rights over raising their children including those people with whom their children should associate.

In the *Blakely vs. Blakely* Illinois case the parents wanted to break off their children's relations with the grandparents because they, the parents, quarreled with the grandparents. It seems obvious that if these parents had banned contact between their children and a neighbor or a neighborhood shopkeeper, for example, this decision would be considered as part of the parents' prerogative or privilege with no means of appeal. Parents have the prerogative to control their children's social contacts to avoid all associates that they deem unwanted, dangerous, or harmful for their children. Even if we, the public, had doubts about the appropriateness of such a decision or the extent to which it fulfills the BIP, we could not raise a convincing argument to render this decision illegitimate. However, the parents in the above case denied the relations between the children and their grandparents even though they themselves admitted that the relationship was meaningful for the children as well as for the grandparents. This kind of harsh decision needs to be closely examined as it strikes at the practice or at least convention of grandparents' visitation, which had been comprehensively acknowledged both socially and legally.

Of course, grandparents' visitation rights are not equal to those of divorced parents. In cases of divorce, there is no doubt that both parents have fundamental and undeniable rights to see their children, and the custodial parent has to respect the rights of the other parent. But in our case, grandparents could not claim visitation rights like divorced parents. Since both parents are competent adults who live together and want their children to stay with them, the grandparents had no executorship or custodianship rights. At most, the grandparents could, perhaps, be considered a "third party," since they still are a significant and meaningful part of the extended family.

In any event, even if relations between the grandparents and the children's parents became acrimonious, the grandparents could still ascribe a moral right to "habit" as a reason to visit their children. This is in line with the theory that holds "that the moral life is primarily a habit of affection and behavior."[7] According to this viewpoint, habit can still establish moral rights and especially in cases where the fulfillment of these rights involves joy and happiness to different family members. When people are accustomed to a certain lifestyle and they benefit from

---

[7] Ross Jacob Joshua. *The Virtue of the Family*. Free press, New York, 1994, p. 93. Ross ascribes this theory to Michael Oakeshott.

continuing that lifestyle, they have established the right to continue their habit without interference (except, of course, if they do something that cancels their entitlement for this right). Any demand for preventing or halting a habit (and in our case, the children–grandparents relations) requires justification. Banning the above grandparents from visiting their beloved grandchildren might be considered a violation of their established moral right, and thus requires very convincing arguments and justifications. We assume that before the quarrel, when the relations between the parents and the grandparents were satisfactory, they enjoyed the benefits of the extended family including relations of love and affection between the children and the grandparents. These relations existed under the parents' awareness and consent.

We must remember that the change in the intrafamilial relations did not result from a clash or hostility between the grandparents and the grandchildren but between the grandparents and the parents. The parents did not argue at any point that the grandparents threatened or negatively influenced the grandchildren, or that the relations between the grandchildren and the grandparents were flawed in any way. The existence of such an argument could, indeed, justify the banning of the grandparents from seeing their grandchildren, due to the desire to protect the children or even from the ordinary claim for children's rights. In such a case, the principle of the child's best interest should unequivocally ban these relations. However, none of this was argued by the parents. All the parents claimed in this context was that since they are the ultimate authority to decide with whom their children will associate, it was their prerogative to ban contact between the children and the grandparents without any overt reason. In fact, their real purpose was to punish the grandparents with whom they had quarreled. The court had to decide whether this right overrides both the rights of the children and the grandparents to maintain their mutual relations, which were established by the previous existence of the extended family relations (which were confirmed by the parents), and to uphold their part of familial relations. When it became clear that there was no risk to the children's safety, and there was no fear that the grandparents will incite the children against their parents, the court decided that the visitation rights of the grandparents were still valid even though the parents were no longer an active part of those relations anymore, and, in fact, objected to their maintenance.

Thus, if there had not been contact between the grandparents and the grandchildren before, and the grandparents wanted the right to see their grandchildren after a long separation with the parents, their claim to see the grandchildren might not be upheld. In that case it could be argued that the extended lack of contact between parents and grandparents constitutes the cancelation of the extended family framework, including the grandparents' separation from the children. Of course, the parents always have the prerogative to allow the grandparents to see their grandchildren, but only on the basis of generosity or supererogation from their side—not as their duty. Since the dispute between the parents and the grandparents occurred long after the establishment of close relations between the grandchildren and the grandparents, their mutual rights to see each other were established and valid. And this right exists even when there are no relations between the parents and the grandparents anymore.

Was the court's verdict a violation of the parents' right to raise their children according to their preferences or desires? I believe not: the decision does not undermine the parents' authority, at least not from the children's point of view. It does not change the way they are educated nor does it impinge on the parents' custodial rights. The court's decision only restricts their authority in a case where the fulfillment of their custodial rights, harms or causes suffering to others. While the court does not ignore the parents' rights to make important decisions for their children, in effect the court is saying that parents still cannot control every aspect of the extended family relation regarding their children. In some sense the court verdict confirmed Jeffrey Blustein's approach, which claims that the duties parents have are prior to any rights they may have. Blustein cites seven points, which are fundamental for determining the parents' status with regard to their relations with their children.

1. "Parents have duties to care for their children (such duties being based on their liberty-right to do so).
2. They have a limited right to a certain latitude in defending and managing childcare, such latitude being necessary for the child's healthy psychological development.
3. They have a right to their children's cooperation and obedience.
4. They have a right to do more than their simple obligation to their children, subject only to considerations of social justice.
5. They have a right to expect honor and respect from their children in return for their care and devotion.
6. They have a right to satisfy their own individual interests, such as privacy, and to engage in non-childcaring activities, provided only that this does not interfere with the satisfactory performance of their child-caring duties.
7. They have no rights other than these."[8]

None of the points above include parental rights to deny relations between their children and grandparents. They cannot deny their children the enjoyment of grandparenthood relations. It is generally assumed that the obligations of adults towards their parents are fully voluntary, and there are no legal claims both sides can raise against each other. Thus, if no love or affection exists between the parents and the grandparents, the parents can claim that no familial relations

---

[8] This quotation is taken from Ross Jacob Joshua. ibid, p. 141. He refers to Blustein approach as presented in: Blustein Jeffrey. *Parents and Children: The Ethics of the Family*, Oxford University Press, New York, 1982.

exist between them and the grandparents.⁹ However, they cannot claim this with regard to the relations between their children and the grandparents, since mutual love and affection still exist between the children and the grandparents, and the parents' obligations to their children requires that they not deny them the benefits of grandparenthood relations and love.

In fact, what the court did was to prevent the use of children as means for settling accounts between their parents and their grandparents. The court made it clear that they would not allow the use of children as ammunition in clashes between adults. By doing so, the court protected both the visitation rights of the grandparents, and more importantly, the children's immunity against being manipulated by adults due to their ultimate dependence on those adults. The court's message was that children should not be dragged into adult disputes and their best interests should be fully protected even in situations of crisis in the extended family's relation. Another ugly example for using children as weapons in clashes between adults, or even between hostile states, is evident in the next example.

CASE 2

This example is the well-known case of Elian Gonzalez, the 5-year-old Cuban child who was rescued by a fisherman on November 25, 1999 in the Caribbean Sea not far from the US shore and brought to the USA. Elian and his mother Elizabet Brotons had fled Cuba and tried to enter Florida illegally in order to join the large Cuban community in Miami. The ship on which they sailed, sank on November 23rd. As a result, the mother died and only little Elian and three others survived. Elian was taken to relatives of his mother in Miami after spending 2 days alone in shark-infested waters. The American authorities, which usually prevent immigrants from entering the US, allowed the child to stay in the US.

The boy's father was Juan Miguel Gonzalez who was divorced from the boy's mother Elizabet and lived in Cuba. He had not given his permission to the mother to take their mutual son to the United States with her new spouse, and he applied to get his son back. He argued that even without dealing with the alleged kidnapping of his son, he was now the child's only legal guardian after the death of the mother. The moral basis for his claim rests on two comprehensively accepted assumptions regarding customary traditions of family relations, which Jacob Joshua Ross believes to be central foundations of the moral relations within family. "(1) The

---

[9] On the Parents–Child Relations After maturity, see Winfield Richard Dien. *The Just Family*. The State University of New York Press, Albany, 1998. p. 157. He says there "Strictly speaking, however, once children have grown up and established an independence of their own, the bond between them and their parents is no longer juridically a family relation."

family serves as society's normal legitimate unit of reproduction and nurture, which means that children and their upbringings serve as its basis ... (2) Nurture belongs, in the first instance to procreators, which is why parents have both special duties and special rights regarding the nurture and upbringing of their children."[10] In the absence of the mother, these rights are exclusively held by the father, and Juan Miguel Gonzalez asked the American legal system to assist in the realization of these rights. He could simply argue that "independently of how it goes for the child, a natural parent can lay claim to be the guardian of her own offspring."[11] This alone should suffice to validate his claim to have his child back.

However, this problem escalated into a political conflict between the United States superpower that dominates world politics, and Cuba, its small boycotted and hated neighbor. The mother's Floridian family assumed responsibility for the child and asked the American authorities to refuse to return little Elian to his father in Cuba. Their main claim was that in rich USA, the child would have greater chances of being raised under better socio-economic conditions than in underprivileged and disadvantaged Cuba. They argued that the child's welfare and best interests require that be raised in the United States rather than in a Third World country, particularly in light of the fact that in Cuba he would be raised in a single-parent family.

There is no doubt that the main linchpin of this conflict was political. If a child had been illegally smuggled under similar circumstances to another Third World country rather than the United States, no one would have dare to challenge the father's exclusive custodianship over his son. The use of such a very young child as an instrument in a dirty political game is nefarious according to any criteria, and I will not deal with this aspect of the conflict here. Elian was taken from Cuba without the father's consent, something which violates The Hague Convention on the Civil Aspects of International Child Abduction that was signed at The Hague on October 25, 1980. Article 1 of the convention states that:

a) "to secure the prompt return of children wrongfully removed to or retained in any Contracting State; and
b) to ensure that rights of custody and of access under the law of one Contracting State are effectively respected in other Contracting States."[12]

Article 8 of this convention demands that "Any person, institution or other body claiming that a child has been removed or retained in breach of custody rights may apply either to the Central Authority of the child's habitual residence or to the Central Authority of any other Contracting State for assistance in securing the

---

[10] Ross Jacob Joshua. *The Virtue of the Family*. Free press, New York, 1994, p. 115.
[11] Archard David William. *Children, Family and the State*. Ashgate, Aldershot, 2003, p. 84.
[12] http://patriot.net/~crouch/haguetext.html.

return of the child."[13] In any event, if this story only involved child kidnapping there is no question that it was a simple criminal act that should be legally resolved by the return of the kidnapped child to his natural parent. But I want to address a different aspect of the story. I want to examine the arguments of the child's relatives about his future welfare in a developed country, that is, the claim that he was more likely to be raised under conditions of poverty in Cuba than in the United States. If we would ignore the hijacking background of the story, the socio-economic claim might be enough to tip the scale towards leaving the child in the USA.

Such a claim requires close examination, since the child's best interests should seriously be considered and protected by society. The big question is what should be the weight of "best interests" in the current conflict vis-a-vis the father's custodianship rights over his son.

In cases where both parents fight for custodianship of their mutual child, the weight we ascribe to the claims of each side is more or less the same. Current social and familial structure and norms tips the scale a bit to the mother's side, particularly when we deal with very young children, and definitely when we deal with babies. However, in this case, the father's claim is not challenged by a claim from the mother or an equally close party, but only from the mother's relatives who had not even known Elian before the disaster. In such a case, we can decisively say that the father's right to raise his child under the current circumstances (where there is no mother) is absolute. No claim about the child's welfare can override the absolute right of the father.

We still need to address a few additional questions, such as the child's judgment. Are Elian's views relevant for the resolution of the conflict? We usually do not ascribe liberty rights to children of Elian's age. We assume that "a child, unlike an adult, simply lacks the ability to make considered choices."[14] This issue is more problematic in the case of Elian Gonzalez because, in addition the fact that he was exposed to manipulations of the mother's relatives, he was too young to understand the principal debate regarding custodianship—that confiscating parental rights from a natural parent (who did not harm the child in any way) would entail dangerous and long-term ramifications. He cannot even understand or appreciate the father's situation of losing a child due to political conflict between states. In such a case, the child's opinion should not receive significant weight when making a decision regarding his custody.

Another issue, requiring a close look, is whether the fact that he would be raised in better material conditions by relatives in the United States, would ensure his happiness and well-being over being raised under worse material conditions by his biological father in Cuba. In this case, the fact that he already lost one

---

[13] ibid.
[14] Archard David William. ibid, p. 33.

parent tips the scales in favor of his return to his father. Being happy, particularly after such a disaster, is not necessarily synonymous with better material conditions, and the child might well be happier with the father. It should also be mentioned that the public health services in Cuba are much better than most private services in the USA, and new emigrants such as the mother's relatives usually have inferior health insurance that would apply to their new family member as well.

Nevertheless, one of the most important questions in this debate is whether the better material life conditions in First World countries can function as a consideration that overrides the parental rights of people from Third World countries. A positive answer to this question would constitute a dangerous precedent for using Third World parents as a sort of "womb for hire" for the more affluent First World, and would subordinate familial integrity to the parents' economic status. It is not coincidental that the father's Cuban relatives supported his stand while the Cuban Miami community supported the stand of the mother who had tried to immigrate to the United States. All of them realized that what was at stake was not only a custodial issue but the status of claims of people from Third World countries to raise their children, vis-à-vis those of the First World. A denial of the father's claims to get his son back, under the excuse that he is much poorer than the mother's relatives, might be used for justifying the abduction of children from Third World countries to be raised by wealthier First World parents—or even the abduction of children from disadvantaged homes within First World countries to be adopted by wealthy families in upper-class neighborhoods. Such dangerous considerations totally distort our concepts about parenthood and family, and accepting them would entail the very loss of our humanism. Economic advantage should not be an argument for confiscating custodianship rights from natural parents. Such considerations may only be justified under extreme conditions, such as when the natural family is starving and cannot provide even minimal nutrition to the child, but this was certainly not the case with Juan Miguel Gonzalez who could feed his son in Cuba.

Fortunately, on Wednesday, June 28 2000, the US Supreme Court reached the verdict to restore Elian to his father and return with him to Cuba. This what eventually transpired, though the relatives resisted violently and the authorities had to raid the family's house to force them to return Elian Gonzalez back to his father. Attorney General Janet Reno issued the following statement in response to the court's verdict:

> "I am very pleased that the Supreme Court has declined to review the case of Elian Gonzalez. The law has provided a process, and this little boy now knows that he can remain with his father. All involved have had an opportunity to make their case—all the way to the highest court in the land. I hope that everyone will accept the

Supreme Court's decision and join me in wishing this family, and this special little boy, well."[15]

The main consideration of the court was that they could not cooperate with the crime of boat hijacking. But this is only one of a constellation of reasons that finally determined that decision. I believe that the final decision made by the American Court was in accord with the United Nations Convention on the Rights of the Child, adopted and opened for signature, ratification and accession by General Assembly resolution 44/25 of November 20, 1989. In Article 3, (1), this convention demands that "in all actions concerning children, whether undertaken by public or private social welfare institutions, courts of law, administrative authorities or legislative bodies, the best interests of the child shall be a primary consideration."[16] This is also the demand of the British Children Act 1989 (c.41), which stipulates in paragraph 1 (1) that When a court determines any question with respect to (a) the upbringing of a child; ... the child's welfare shall be the court's paramount consideration."[17] It is not only the fact that the father in the only one who possesses parental rights, but that it is the child's best interests to grow up with his father in the place he was born, than with relatives that he first met only after the disaster of losing his mother.

This belief is based on the comprehensive understanding, cited by David William Archard, that "the presumption that natural—that is, causal—parents should be allowed to care for their own children is the predominant understanding of parenthood in modern society."[18] He brings some justifications for this assumption, two of which are worth quoting:

> "First, natural parents are the best suited to care for their own children. Biologically disposed them to give love to their offspring—a love that is often unconditional and self-sacrificial. Second, being raised by one's own natural parents contributes appropriately to the identity and self-image of the maturing child. The health development of any person requires the successful acquisition of a positive and stable self-image. The fact that the child and parent share heredity and, in consequence, are significantly alike in respect of appearance, skill, and general abilities greatly facilitates this. In turn, such mutual identification makes possible and fosters the deep, reciprocated affection the characterizes the best parent-child relations."[19]

Of course, not all parents treat their natural children with "unconditional and self-sacrificial" love, but they are much more likely to so than other, unrelated adults. Thus when we need to decide what is the child's best interest, we may assume the most plausible assumption—the natural parents.

---

[15] http://www.usdoj.gov/opa/pr/2000/June/373ag.htm.
[16] http://www.unhchr.ch/html/menu3/b/k2crc.htm.
[17] http://www.hmso.gov.uk/acts/acts1989/Ukpga_19890041_en_2.htm#mdiv1.
[18] Archard David William. ibid, p. 84.
[19] ibid, p. 84–85.

CASE 3

About 7% of American couples encounter the thorny problem of infertility.[20] Reproduction technology offers hope to these couples, but also additional challenges as new techniques create situations where, theoretically, a baby may have "as many as five different parents': a genetic mother and father, a gestating mother, and a rearing mother and father."[21] In the example below I concentrate on one aspect of the relations between those who were born as a result of sperm or egg donation, with those who made the donation many years before.

In January 2004, the British Government decided "to end anonymity for egg and sperm donors... despite strong opposition from specialist doctors in fertility clinics."[22] Previously, children born as a result of egg or sperm donation in Britain had access only to a very limited range of non-identifying characteristics of the donors such as eye color, hair color, height, etc. They had also access to the medical history of the donor, but without any access to his/her identity. The new law, which went into effect on Friday, April 1, 2005, states that "any children conceived after that date will be able to learn their donor's identity when they turn 18."[23] Hetty Crist, a spokeswoman for the Department of Health in London, justified this new decision by what she considered as the right of the children to have this information.[24] Of course, this decision will be relevant only for children whose fertilization happened after the law came into force. Earlier donors will still have to give their consent to be identified by any biological offspring interested in such information, and can choose to retain their anonymity if they wish.

The objections to this new decision came both from doctors and donors. The doctors fear that the new law will end altruistic donation of eggs and egg-sharing schemes. The problem with sperm donation will not be as acute, since today's clinics can solve almost all the problems of men with low fertility. However, the problem with egg donation is likely to be acute. As it is, couples wait a long time for egg donations and demand for these donations is continuously on the rise— while the new law is likely to discourage women from donating eggs in the future. The opponents of the new law brought examples from Sweden and New Zealand, where such laws were initiated and the donation rates declined soon after. The donation rates stabilized again after a year, but on lower levels. In any event,

---

[20] This detail is taken from Murray Thomas H. *The Worth of a Child*. University of California press, Berkeley, 1996, p. 16. Murray brings this number from: Rosenthal Miriam B. "Psychiatric Aspects of Infertility and Assisted Reproductive Technology." In: *Psychological Issues in Infertility*. Vol. 4, No. 3, 1993, p. 471.
[21] Murray Thomas H. *The Worth of a Child*. University of California press, Berkeley, 1996, p. 16.
[22] http://www.intendedparents.com/News/egg_and_sperm_donors_to_be_named.html.
[23] http://www.csmonitor.com/2005/0330/p11s02-lifp.html.
[24] ibid.

there is still "reason to suspect that donors would not participate in donation if anonymity were not guaranteed."[25]

The legal situation regarding the anonymity of donors differs from county to country. Switzerland accepts only donors who are willing to be identified, while France insists on compulsory anonymity of its donors. The United States takes a middle road where donors decide whether to be anonymous or not. In fact, there are sperm banks in the USA such as Pacific Reproductive Services in San Francisco that pays higher fees to donors who are willing to be known than it does to those who choose to remain anonymous. Finally, there is no issue of anonymity in Italy because Italy does not allow donor insemination at all. Thus, it seems that there is no consensus regarding the legitimacy of donor anonymity, no accepted or dominant norm on the issue, and no way to predict the possible effect the new law will have on donation rates.

The donor-anonymity conundrum demonstrates the conflict between the children's desire to receive information about their biological parents, and the donors' rights to privacy. Even if the donors' would not be accountable to any financial, emotional, or other claim made by the biological offspring, they still maintain their right not to have their identity revealed. Their immunity from any financial or emotional claim will be fully respected, since we believe that "the mutual rights and obligations of parents and children depended morally not on the biological facts alone, but also and more particularly upon the nurturing and care the natural parents provided. If he or she had not has any part in the nurturing of the child, the child could not be expected to look upon him as a father or her as a mother."[26] It is this approach that is the moral basis of the institution of adoption, when the authorities give the adoptive parents the same rights and obligations of parenthood that are waived by the biological parents.

Initially when this law was prepared, donors were concerned about their privacy and terrified by the possibility of being revealed by children who were born from previous sperm donations. This fear (of losing anonymity) was significantly reduced by the fact that the law is only relevant for future and not past donors. All donors of the 18,000 children who were born after 1991 retain the option of remaining anonymous. (It was in 1991 that the British Human Fertilization and Embryology Authority was established, which regulated all the authorized fertility treatment clinics to document the details of every sperm or egg donor).

Nevertheless, this new law signals a change in the current practice of sperm or egg donations when compared to other organ donations. Anyone who receives any other organ—such as a kidney or heart—cannot insist on revealing the identity of the donor, if the donor wishes to remain anonymous. The new law that "ends

---

[25] O'Donovan Katherine. "What Shall We Tell the Children? Reflections on Children's Perspectives and the Reproduction Revolution". In: Lee Robert and Morgan Derek. *Birthrights*. Routledge, London, 1989, p. 106.

[26] Ross Jacob Joshua. *The Virtue of the Family*. Free press, New York, 1994, p. 115.

anonymity for egg and sperm donors" seems to acknowledge or affirm that there is some kind of biological, even parental, relationship between donor and child—otherwise, the children born of donors would have no reason to demand the identity of the donors any more than of kidney or heart donors, except for genetic reasons (explained below). Thus according to the new law, sperm, or egg donation endows a special status on the offspring that entitles that offspring with some valid claims which override the donor's right to anonymity. And this status hints to a sort of familial relations, or at least familial connection.

Such an acknowledgment of the child's status towards the donor dramatically changes the current understanding of the practice of sperm and egg donation, both in the eyes of the donors and the eyes of the fertilization authorities. Until now it was understood that except for the act of donation, there would be no connection or relationship of any kind between the donor and the offspring, and certainly not biological parenthood. It was a given that sperm and egg donations did not create relations between the donors and those who receive the donation, any more than other voluntary donations of other organs. And was accepted that the donor could remain anonymous. Now, the new law demands that the donor be willing to assume the risk that he or she may receive a sort of quasi-parental status sometime after 18 years, to an offspring he never met before. Since the offspring can ask to know who was the donor at any point after reaching the age of 18, that means that the donor can be approached by the offspring at any point in his or her life as well. This is a completely different situation than simply agreeing to assist someone else to have children as part of one link in a long chain of actions including a physician and medical intervention. Thus the new status of the donor, and the concept of sperm and egg donation as a whole, must be conceptually reconsidered and remodeled.

A more serious complication is that donors might demand to know the identity of the children who were born as a result of their donation and might even submit a claim for some extent of parental status with regard to these offspring when these children reach the age of 18. Such a demand would place the offspring in impossible and unbearable situation, as well as the parents who raise the child conceived of sperm donation and might then be requested to share at least part of their parenthood with a third party. These are much more critical problems that the issue of donor privacy. Thus the law should be clarified to set forth the rights of all parties clearly, to avoid painful confrontations in the future.

To avoid the latter problem of donors making parental claims, we must adopt the offspring–donor asymmetric relationship that is accepted regarding adopted children and their biological parents. In the case of adoption, children can ask to know details about their biological parents but not vice versa (which means that biological parents who waive their custodial rights to give up their children to adoption, cannot ask for information about the children who were adopted by other parents). This asymmetry should be preserved in the new law regarding sperm–egg donation as well.

There is, however, a case to be made in favor of the right of offspring to know the identity of their biological parents, and that is in relation to genetic-related diseases or problems. The genetic similarity of biological parents and their children can sometimes save life, such as in the case of bone-marrow transplants when first-order family relatives have the greatest chances of having genetically matched bone marrow. There were cases when adopted children contacted their biological parents for this specific reason; on such occasions, the information about parental identity is invaluable. However, it is possible to solve this problem by invoking the asymmetry typical of adopted children: the children have the right to seek the sperm or egg donor, but not the reverse. If the donor would need the transplant (and not the reverse), then anonymity should be preserved while the authorities should try to convince the offspring to be tested for bone marrow donation, without revealing the identity of the donor/quasi-parent. In short, donor anonymity should strictly ensured under all circumstances.

To summarize this chapter I can only say that new procreation technologies, and their ramifications on modern forms of parenthood, must be closely examined to avoid the danger of destroying the family framework as a whole. The family is the foundation of any society and its collapse might signal the demise of the social framework as a whole. Thus, those who should resolve conflicts regarding familial relations must be very careful, particularly when children are involved. The next chapter demonstrates the complexity of cases that result from modern advanced procreation technology.

Chapter 9

PROCREATION AFTER DEATH

In the previous chapter we saw that not only that there are several legitimate forms of a family in modern society, but also that the new forms of family create different claims for familial relations. In addition, we have seen that the new procreation technologies that enable various forms of familial relations—such as IVF from donated sperm—bring also new options for having family members.

This chapter deals with such options, which involve claims made by adults to have new family members. These options became possible due to the progress, which have been recently achieved in fertility technologies, besides solving infertility problems, have also changed some of our basic concepts about family life and family relations. For example, the concepts of "reproductive autonomy" or "the right to choose" were originally mainly employed in the context of discussions of using contraception devices and having abortions, but now deal primarily with the right to assisted procreative technologies such as IVF from donated sperm. Onora O'Neill describes this change by saying that "Attention shifted from the problem of controlling unwanted fertility (although the abortion debate lost none of its steam) to that of dealing with unwanted infertility."[1]

We no longer debate the legitimacy of using contraceptive devices, and a large percentage of the public has also accepted the legitimacy of abortion. We can say that the public agenda has turned now to polemics and discussions regarding the limits and constraints of the use of new procreation technologies. For example, while we have already accepted IVF as a legitimate or even standard solution for barren parents, we still decisively reject human reproductive cloning as such a solution. However, between these extremes there are many intermediate cases

---

[1] O'Neill Onora. *Autonomy and Trust in Bioethics*. Cambridge University Press, Cambridge, 2002, p. 57.

where we weigh whether to support or object to requests for using assisted reproduction technologies in order to have biological offspring. In this chapter I will discuss two of such borderline cases: requests to bring a child into the world from sperm taken from a dying relative.

One of the most complicated issues relating to parenthood and family relations is the desire of parents or spouses who have lost a beloved male relative to use the deceased's sperm, in order to create a new generation of that same family. One example of this is of a man who knows that he is dying, and opts to donate his sperm so that his wife can bear his child after his death. Sperm donations may be frozen and ready for use through IVF technology even long after the donor's death. Another option, which has become possible only recently, is the harvesting of a dying man's sperm. This may be relevant after a car accident or severe injury when there is still a window of opportunity for the physician to extract viable sperm prior to death. However, this case differs from the previous one in that the seriously injured or unconscious victim is usually not able to provide informed consent.

In such cases, the moral dilemmas are much more complicated than the technical problems and present very subtle moral considerations to be weighed. Even if we set aside for a moment the issue of consent, the family structure that results from inseminating the man's sperm within the same family will obviously not follow the traditional form of two adults and their children. The variation may involve a skip-over or a confusion of generations (as in the case of grandfather-grandmother-child, or grandfather-mother-child, and so on). However, the central moral problem with regard to such use of procreation technology starts from the fact that it is important to consider children as ends unto themselves, and not as means for any other reason—no matter how understandable it is for people to want to preserve the memory of a dear departed spouse or child. But we know that parents do not only have children for "pure" motives, as "ends into themselves." We have to admit that "children have helped to meet a variety of adults needs, as household workers or as support in old age; emotional needs, for intimacy and affection; and development needs, for maturation, for ripening of the virtues appropriate to adult life,"[2] as Thomas Murray reminds us. And since children are especially vulnerable to exploitation, we have to be very circumspect and prudent about any possible (even unintentional) use of children for the fulfillment of adults' interests or desires. And the most frequent of these desires, and probably the most natural one, is simply to have a child of one's own. In such cases we should consider all the interests and needs of future children, even before they are born or even before they are conceived. Thus, under certain circumstances we can question the legitimacy of bringing children into the world, and even the legitimacy of the fetus' conception. Such a discussion deals with the interests of not yet extant entities whose eligibility to have interests or needs is doubtful,

---

[2] Murray Thomas H. *The Worth of a Child*. University of California Press, Berkeley, 1996, p. 8.

making the entire discussion very problematic/complex despite its rationale. The coming paragraphs outline two examples of such considerations.

CASE 1

Adults often have a strong desire to have a child, reflecting David William Archard's word that: "it is undoubtedly true that having children engages some of the deepest and most central values in a person's life. It also seems true that what matters most to people is having their own children."[3] He also reminds us that "The desire of an adult to have her own child is inseparably bound up with the adult's understanding of what gives her own life significance and value."[4] However, Archard stresses that, extremely powerful as that desire may be, it must not be selfish and the needs of the future child must also be taken into consideration.

The first example of the two is a request made by a young childless widow to procreate from her dead husband's sperm. This sperm was harvested from his body just before his death, while clinically he was actually dead. Even though circumstances dictated that the husband could not give explicit consent before his sudden death, his parents and the extended family of both spouses (of the dead husband and the widow) encouraged and supported this initiative as well; they even appealed to the courts to approve this request. The Israeli legal authorities were won over by the show of willingness and support of the widow and of both extended families, allowing her to procreate from the sperm that had been harvested. In the context of this case we could responsibly say that "the freedom to pursue parenthood is one of the most important expressions of individual liberty."[5]

The widow's request was an explicit expression of her desire to realize her procreation freedom and have a child as a single parent. And since the single-parent family has been widely accepted in modern society, there are not many objections one could raise to this request. In fact, the single-parent family phenomenon has grown immensely over the last three decades. Jon Bernardes informs us that "In 1994–5, 22 percent of all families with dependent children in the UK were lone parent families, compared with 7 percent in 1972,"[6] an increase of 300% within 22 years. The causes of this phenomenon are the higher rates of divorce, but also due to the greater legitimacy of unmarried pregnancy. If we examine the rates of children who were born to unmarried mothers in Sweden during the third quarter of the 20th century, for example, we find that these rates had increased from 10.2% of all births in 1958 to 35.9% in 1978 (again, this rate more than tripled

---

[3] Archard David William. *Children, Family and the State.* Ashgate, Aldershot, 2003, p. 79. The emphasis is in the original.
[4] Ibid.
[5] Murray Thomas H. Ibid, p. 19.
[6] Bernardes Jon. *Family Studies.* Routledge, New York, 1977, p. 18.

itself within 20 years).[7] It is safe to assume that these rates are even higher now. There is no doubt that the single-parent family paradigm has been accepted in most Western societies.

It is true that the lone-parent family features one parent, usually the mother, who raises a child without a father. Conventional wisdom has long discarded the concept upheld by such people as Warren T. Reich or Paul Ramsey who maintained that "the preferred moral option is that governmentally sponsored in vitro fertilization with embryo transfer be restricted to the marital context—that is, that it be restricted to married couples who are attempting to bear their own child by use of their own gametes."[8] The assumption upon which they based their demand, that the "covenant of love (marriage) provides essential means and assurances for parenting and for nurturing of the child,"[9] as well as their attitude which claims that the sexual-physical-relational exclusivity and unity of the marriage covenant "make marriage a source of the stability that responsible parenthood requires,"[10] has been refuted long ago from all possible aspects. We find married couples who proved themselves to be irresponsible parents, as well as unmarried couples (including same-sex couples) or singles who proved to be excellent parents. Thus, we do not restrict the right to fertility treatment to any specific form of the family, but we do set criterion regarding the competence of those who request assisted procreation treatments so that they can become parents.

In this context we cite Onora O'Neill who states that "To restrict access to reproductive technologies to those who are fit, capable and committed to being at least adequate parents is no more discriminatory than restricting fostering or adoption to those with adequate health, capacities and commitment."[11] She does not restrict assisted reproductive technologies only to married couples, but insists that the criterion for those who seek advanced reproduction technologies to have a child, should be the same as for people who apply to adopt a child, for example. The test should be "whether they are capable of and committed to being present and active for a child across a very large number of years."[12] Since there is no reason to believe that the widow does not fulfill these requirements, there is no reason to prevent her from using these treatments. In addition, the fact that the deceased husband's parents—as the future grandparents of the child, and thus as having interests in the child's welfare—support

---

[7] Trost Jan. "Changing Family and Changing Soseity." In: Trost Jan (ed.), *The Family in Change*. International Library, Vasteras, Sweden, 1980, p. 13.

[8] Reich Warren T. "In Vitro Fertilization and Embryo Transfer: Public Policy and Ethics." In: Teichler-Zallen Doris and Clements Colleen D. (eds.), *Science and Morality*. Lexington Book, DC, Heath and Company, Lexington, 1982. p. 113.

[9] Ibid.

[10] Ibid, p. 114.

[11] O'Neill Onora. Ibid, p. 64.

[12] Ibid, p. 65.

her request, provides another reason to believe that the widow is fully competent to raise a child. And this endows her with the right to be assisted by that technology.

We have long internalized the conception of the family as what Lainie Friedman Ross calls "an intimate group," while the exact make-up or composition of the family is secondary to this concept of "an intimate group ... in which members derive benefit if not identity from membership ... Members make compromises for each other in order to promote the goals of the group in addition to their individual goals. It is a group in which members try to accommodate one another's needs rather than disengage."[13] What is unique in the intimate group of a family with children, is the parents' obligations to secure at least the minimum for each child (or minor) in the family.

Thus the family created by the widow from the deceased husband's sperm—that is, the widow and her biological child conceived from the deceased husband's sperm—definitely would fulfill this criterion of an intimate family. Her request, therefore, should be granted even though she intends to raise her child alone, without a father. Her situation is likely to be easier than the usual form of a lone-parent family since she will have the support and assistance of her husband's extended family in addition to her own. And the child will have the benefit of two pairs of loving grandparents, not one, to the benefit of all involved.

Nevertheless, there are some marginal issues that should be addressed. One is the unraveling of parenthood from the sexual act. However, this phenomenon has become such a frequent and almost commonplace procedure that it no longer affects the parent–child relationship or kinship in any way. Parenthood is no longer deemed a kind of station that follows only from one's own procreative activity. We assume that "parenthood involves more than just procreation; it entails entering into a relationship and assuming a responsibility for the rearing of offspring regardless of their origin."[14] Thus the woman's procreation without the sexual act is not a relevant ethical consideration, equal to the case of a woman using a surrogate mother to have a child. The separation of parenthood from the sexual act has become almost "a non issue," at least from the ethical point of view.

Another issue in the present case is the concern that the child will be treated as his father's monument, something which might undermine his autonomy, self-determination, and independent personality (let us assume we are talking about a boy). Here it is the mother's responsibility to avoid this pitfall and instead, raise the son as if the fertilization occurred naturally or before the father's death. The special and tragic circumstances which caused the need for assisted procreation treatment should not affect the mother's obligation to ensure and foster the child's

---

[13] Friedman Ross Lainie. *Children, Families and Health Care Decision Making.* Oxford University Press, Oxford, 2002, p. 34.
[14] Reich Warren T. Ibid, p. 111.

self identity and individual autonomy in any way. Being a parent in this sense includes the usual duty to nurture the child's personality, and this means the opportunity to allow him to have his own identity. In any event, this would not be a reason to bar special fertility treatment from the woman, It only imposes additional responsibility on the mother to ensure that the child be treated as an "end" and not a means or in this case, as a monument to the dead father. The main reason for the son's existence should remain his own personhood, and not the father's memory or ancestry. A competent parent can ensure this. We believe that "All parenthood exists as a balance between fulfillment of parental hopes and values and the individual flowering of the actual child in his or her own direction,"[15] as Dena Davis reminds us. The above widow should be aware of the fact that the child will be a person, an individual, a subject, and thus an end unto himself. The fact that he never met his father, should not affect his independence and development from this sorrowful context. A similar case is that of parents who, after losing a child, decide to bring a new child into the world in order to ameliorate their agony and help them in dealing with their bereavement. They bear the same responsibility to ensure that the new child will not function as his brother or sister's monument, but be considered as an end unto himself.

Even if we assume that the widow's decision to have a child from the deceased husband's sperm was not made for the sake of the child but for her own sake and for the sake of perpetuating the husband, she has to make sure that the child remains an end unto herself (let's assume it is a girl). And being a good parent requires her to balance between having this child for her own sake "and being open to the moral reality that the child will exist for *her* own sake, with her own talents and weaknesses, propensities and interests, and with her own life to make."[16] The mother who is careful to nurture the child properly, would respect the child's right to develop her own self identity and personhood.

The only remaining problematic issue in the current case is the absence of the husband's consent, though under the circumstances we could possibly presuppose his tacit consent. (Since the wife wanted the biological child, it could be assumed that she knew the husband's general feelings about having children.) The fact that the husband's parents and extended family supported this request can perhaps also be understood as an acknowledgment or affirmation of this consent. Conversely we can say that since the fulfillment of this request will not raise any future claims against the deceased husband, and since the only party who might be accountable for any future claim is the grandparents, their consent as well as their support of the widow's request can be understood as substitutional consent to that of the dead father. Thus the decision on that request was not morally problematic.

---

[15] Davis Dena. *Genetic Dilemmas*. Routledge, New York, 2001, p. 34.
[16] Ibid.

CASE 2

This second example presents an issue that is mainly problematic on moral grounds, not legal ones.

In this second case, an elderly couple had a son who was very sick with a terminal illness. The dying son willingly donated sperm to fulfill his parents' request to transplant this sperm in a surrogate mother's womb; the surrogate would then give them the child after birth. Since this was the original reason for the donation—that is, the son consented to this scenario—the Israeli legal authorities allowed them to use this sperm for the purpose of having a biological grandchild from their dead son's sperm.

If we were holding this discussion a number of years ago, we might have had to address the problematics of the surrogacy issue as well. However, surrogacy today is now more accepted as one of the new options that have become available to overcome infertility problems. It is also widely accepted that the surrogate mother must undergo a rigorous selection process to minimize problems later (such as the surrogate deciding that she wants to keep the baby herself). Although there are, occasionally, clashes and conflicts between the surrogate mother and those adopting the baby, these not much more frequent than in adoption when, for example, the biological mother suddenly regrets giving up the baby for adoption and sues to get back her biological child. Many people believe that the concept of adoption has helped us to internalize surrogacy as a permissible option within the options crated by the new procreation technologies. Rosamond Rhodes even holds that adoption "should count as *prima facie* evidence for continuing to regard surrogacy as ethically accepted."[17] Like other forms of collaborative reproduction, such as sperm and egg donations, surrogacy has been woven into our social fabric for the benefit of all affected parties. So, the surrogacy issue is minor in the current case and I will skip this in my discussion.

The primary objections I raise to this case is the advanced age of the couple and the likelihood that they may not live long enough to raise their grandchild until the child reaches independence. The age differences between them and their grandchild and average life expectancy tables in the Western world, predict such a possibility. Even if they physically survive (if they exceed the average life expectancy) their physical constitution is likely to be impaired so that not only would they be incapable of taking care of their child properly, but they might need the child to take care of them when he reaches his teenage years. In short, the fear is that the most fundamental rights of this child, what Joel Feinberg calls *The Child's Right to an Open Future*,[18] will be violated as a result of the grandparents' advanced age.

---

[17] Rhodes Rosamond. "Reproduction, Abortion and Rights." In: Thomasma David C. and Kushner Thomasine (eds.), *Birth to Death*. Cambridge University Press, Cambridge, 1996, p. 69.
[18] See Feinberg Joel. *Freedom & Fulfillment*. Princeton University Press, 1992, pp. 76–97.

Most countries prefer not to let older couples adopt children and when they do allow it, there is usually a reason; for example, if the child is older or has severe health problems, it may be difficult to find younger adoptive parents. In most adoption cases, the child's life prospects before the adoption are usually much lower that those after the adoption, even if the adopting couple is fairly elderly. It is a case of being the lesser of the two evils. The obligation to ensure that children receive the greatest possible life opportunities (part of which is imposed on the state as the children's *Parens Patriae*[19]) requires us to make as many opportunities as possible available to each child. Because a child who is given up to adoption usually has very few life prospects before the adoption (as the biological mother does not want to raise the child), the act of adoption itself opens many more opportunities for this child.

But in the current case, where the child has not yet been conceived, we have to consider whether by allowing elderly parents to produce a child with procreative technologies, we are not limiting that child's future opportunities. We recall Onora O'Neil's claim that the criteria for assisting people to have children with procreative technologies should be the same as to those who do not need such treatment; thus we compare this case to that of adoption. There are few chances (at least in Israel, but in many other countries as well) that elderly parents are allowed to legally adopt a baby, due to their fairly high age. And in our case there are similar considerations to reject these parents' request.

The child's right to an open future is an aggregate of rights, which are reserved only for children, and do not stem from their physical dependence on their parents (since such rights are also reserved for mentally retarded adults). This comprehensive right includes the things whose preserving is necessary for the appropriate development of the child, so when this child arrives adulthood he/she will be able to realize all the rights which are given to adults in his/her society. However, this right can be violated "in advance," that is, the violation of the right may occur even before the child is in a position to exercise it. Joel Feinberg says the "in advance" violation means that "The violation conduct guarantees *now* that when the child is an autonomous adult, certain key options will already close to him. His right while he is still a child is to have these future options kept open until he is a fully formed, self-determining adult capable of deciding among them."[20] Having old parents might raise a fear that this right might be violated (even unintentionally, but perhaps unavoidably) "in advance," and I will try to explain why.

The unique character of this right is that it requires that the parents impose restrictions on the child in order to make sure that the child does not make poor choices that will ruin his life later. For example, they may have to force him to

---

[19] A doctrine whereby the state takes jurisdiction over a minor living within its border. Usually it is the basis for deciding what state will assume jurisdiction in a child custody case, but it also used for imposing responsibility on the state, to ensures the minors' basic rights within it territory.

[20] Feinberg Joel. Ibid, p. 77. The emphasis is in the original.

have his vaccinations to prevent disease, or compel him to attend elementary school against his will. In the current example, the concern is that the elderly grandparents will not be able to fulfill the child's needs and rights to an open future as they themselves get older and more infirm. Raising children, as any parent knows, often involves great emotional, financial, and even physical effort and many sleepless nights; efforts that are a challenge for young parents, and perhaps insurmountable to older ones.

This case is analogous to the issue of imposing a maximum age limit for a woman's fertility treatment. The immediate consideration is the health of the intended mother and her ability to physically bear and deliver a child safely. However, the additional considerations are similar to our elderly grandparent case. For example, in the UK, elderly applicants for fertility treatment may be rejected through considerations of "the good of the child (especially when he is growing up, when his aged mother may become a tremendous burden, a responsibility, and a source of guilt and embarrassment),"[21] as Mary Warnock explains these reasons. She also explains that "in France it is illegal to give assistance for a postmenopausal woman to conceive, the law being based on the concept of the good of the child, as well as on a widely shared feeling expressed in the statement that 'there is a time to be a mother and a time to be a grandmother', and that such late birth contravene the common sense of law."[22] The rationale for imposing such age limitations is that raising a child is a long-range project of ensuring a child's welfare over many years, thus the parents should be at an age where the odds are in their favor for living long enough, and in good health, to raise the child until independence.

Even though in this case there were no clinical reasons to reject the grandparents' request, since they intended to be assisted by a surrogate mother anyhow, there are still the considerations of the child's future welfare and development.

When a child has elderly parents in poor health, this creates problems for the child despite the fact that minors have no legal obligations to take care of their parents. In addition to the emotion burden, the child as a minor (economically and physically dependent on his parents) still maintains reciprocal relations with parents and this imposes certain duties and obligations on the child (perhaps shopping, cleaning, taking the parents to physicians, etc.) that other children do not have. Or as Jacob Joshua Ross says, the child's duties "of filial piety (honor and respect) to their parents,"[23] starts to include not only emotional obligations but also material and concrete ones. This is something that the grandparents should consider before deciding to function as the child's custodial parents.

Onora O'Neill believes that "The dependence of children provides good reasons to restrict the use of assisted reproductive technologies to those with adequate

---

[21] Warnock Mary. *Making Babies*. Oxford University Press, Oxford, 2002, p. 49.
[22] Ibid, p. 48.
[23] Ross Jacob Joshua. *The Virtue of the Family*. The Free Press, New York, 1994, p. 159.

health and capacities, who have reasonable expectations and intentions of being active and present to bring up the child the aspire to being into the world."[24] By saying this she means that those who "lack the health, capacities and commitment for coping with the demand of parenthood"[25] have no right to be assisted by new reproductive technologies. However, this guideline still does not provide us with a good answer whether those parents, who are healthy but old, would meet the requirements mentioned earlier this chapter, which O'Neill imposes on parents who apply for assisted reproductive technologies. We cannot really say for how long these grandparents would be capable of being present and mainly active for the born child.[26] O'Neill thinks that in the decision whether or not to provide assistance in procreative technology the main question is "whether there are reasonable grounds to think that any child brought into existence can expect to have at least an adequate future, cared for by a 'good enough' family (biological or not) who will present and active for the child across the long run."[27]

Onora O'Neill refers in a short paragraph to the issue of elderly woman who wants to procreate, and says:

> enabling elderly women to have children whom they are unlikely to have the health and years to nurture and well knowingly risks serious difficulties for any child. The fact that other children with younger mothers may also lose their mothers too early in life is not a sufficient argument for choosing to have children vary late in life: such loss is widely seen a grave misfortune, and hardly something for which even minimally responsible aspiring parents would plan.[28]

I would not deny these elderly parents from using novel procreative technologies to produce a grandchild from their dead son's sperm. However, I would attempt to convince them not to raise the child on their own. In this context I agree with Kenneth D. Alpern, who says that "having (one's own) children, in its many senses, is undeniably a fulfilling experience for most people. In the present state of our consciousness and culture, prudential, moral, and public policy decisions should all respect this important good. But the good of having (one's own) children should not be blindly indulged in any of these spheres of evaluation."[29] While the desire to have a child is an "undeniably... fulfilling experience," it must be weighed against the needs of the child to be raised by competent, healthy parents—and such an issue is extremely sensitive when dealing, as in this case,

---

[24] O'Neill Onora. Ibid, p. 63.
[25] Ibid, p. 64.
[26] These criteria are in O'Neill Onora. Ibid, p. 65.
[27] Ibid, pp. 66–67.
[28] Ibid, p. 67.
[29] Alpern Kenneth D. "Genetic Puzzles and Stork Stories: On the Meaning and Significance of Having Children." In: Alpern Kenneth D. (ed.), *The Ethics of Reproductive Technology*. Oxford University Press, New York, 1992, p. 165.

with bereaved parents. My suggestion is to search for alternative, creative ways to fulfill the natural and strong desire to have children. Alpern hints to such a solution when he says, "people can realize the value of a variety of creative interpersonal and intergenerational activities. We can form our lives to be less dependent of a single way of satisfying our desire for the generic good of community, caring and participation in the processes of life."[30] I would propose a good, though imperfect, solution to the parents' desires to preserve their son's posterity: to donate the frozen sperm to a woman who wants to have a child. But instead of functioning only as a surrogate mother, she would actually adopt and raise the child. This woman, however, would agree in advance to maintain a degree of family relations between the child and the biological grandparents, as is the case today in certain kinds of open adoptions. For example, she would agree in advance to allow the grandparents to visit and take the child for trips and visits, she would maintain contact with them by phone and mail, she might even accept financial support from them, or use the biological father's family name for the child. In short, she would grant them quasi-grandparent status.

Such a compromise reduces the fear that the child will be orphaned at an early stage of his/her life, or be required to take care of elderly parents when still a teenager. The biological grandparents would be able to enjoy the kinds of familial relations that real grandparents have with their grandchild without having to function as full-time actual parents. And, of course, they will be consoled by the fact that they have continued their son's legacy. Of course, this solution does not fulfill their desire to raise their grandchild by themselves and they would have to content themselves with only visiting rights, but the child's right to an open future will be better preserved by such a solution or arrangement. Although it is not legally required to take into account the "open future" rights of a yet unborn human being, we can argue that it is morally imperative to do so, certainly according to the moral issues raised by Onora O'Neill.

To sum up this chapter I want to reiterate the point that in the current issues about procreation right "reproductive autonomy includes not only the elimination of involuntary reproduction but also the overcoming of involuntary infertility."[31] The overwhelming desire of people to have children (or as in our second example, grandchildren) sometimes leads them to consider novel means to achieve their reproduction goals, without considering the ethical and social uncertainties involved. Thus, I maintain that it is the obligation of the authorities to step in to ensure that desperate people who cannot be counted on to make reasonable choices will use the advanced reproduction technologies "in responsible, constructive ways that minimize harmful effects on participants and offspring."[32]

---

[30] Ibid.
[31] Morgan Derek. "Surrogacy." In: Lee Robert and Morgan Derek. *Birthrights*. Routledge, London, 1989, p. 78.
[32] Robertson John A. *Children of Choice*. Princeton University Press, Princeton, 1994. p. 3.

However, in most cases we should respect people's procreative liberty and their right to use reproduction technology for having children of their own. It is only in extreme cases that the authorities should be allowed to intervene. Generally, in cases of conflict between different parties, the main consideration that can override procreative liberty is the mother's health. I hold that in addition, considerations of the future child's welfare and "best interests" should have a certain amount of weight in such conflicts. Thus I would raise objections to the grandparents' request to raise their biological grandchild and suggest the alternative of visitation rights to an adoptive family, but if they insist then I would not deny them the right to raise the child, mainly for reasons of mercy. But it will be society's mission to take full responsibility for that child, as well as for the aging grandparents in case of future need. This is the real meaning of membership in solidary society as a society that should support the needy individual as a kind of extended-community family.

Chapter 10

BABIES AS COMMODITIES

The dimensions of global trade between rich and poor countries—certainly a direct consequence of current globalization—have both transcended traditional state boundaries as well as transformed anything and everything into objects of trade and commerce. That human beings have been used as commodities is familiar in human history: witness slavery and prostitution. However, the extension of this regrettable phenomenon to babies is one of the most objectionable aspects of our new global form of life and thought. What started as a generous movement of international adoptions, with well-meaning motivation and intentions, has deteriorated into a capitalistic profit-making venture in which babies are no more than the means of maximizing profits. In this chapter we endeavor to pose, and answer, queries about the ethical implications of the tragic move from adoption to baby commerce.

One of the main reasons for the regrettable phenomenon of baby trading is that "At the present time there are many more people who want to adopt than there are babies available."[1] Even though there are many children who grow up in public custodial institutions, people prefer to adopt babies rather than these older children. David Archard maintains that adopted and foster children "may be much harder to rear" than other children.[2] In addition to the usual difficulties involved in raising children, and of adopted children in particular, children who are adopted at older ages tend to have more problems, both physical and emotional. Sometimes they were maltreated or abused by their biological parents before they were placed in foster or custodial institutions, making it difficult for adoptive parents to bond

---

[1] Tizard Barbara. *Adoption: A Second Chance*. The Free Press, New York, 1977, p. 1.
[2] Archard David. *Children. Rights and Childhood*. Routledge, London, 1993, p. 146.

with them and gain their trust in order to enable them to have happy and healthy childhoods.

Thus it is much easier to rear babies than children, even though it is still more difficult to rear an adopted baby rather than one's biological child. The tension caused by the large demand for babies for adoption and the scarcity of supply leads directly to many moral problems, one of which will be exemplified in Case 1 of this chapter. In any event, the very use of the terms of "supply and demand" has its own moral problems, since they are part of what Margaret Jane Radin calls market rhetoric.[3]

The drastic scarcity of newborn babies available for adoption during the last quarter of the 20th century was caused by the growing demand of mainly infertile couples, but also by singles who were ready for single parenthood and less likely to give their children up to adoption. For example, the UK Office of Population Censuses and Survey show that over 1500 children under 6 months of age were offered for adoption in 1976, while only 472 babies were placed for adoption in 1986.[4] This decline brought the Editor of the British Agencies for Adoption and Fostering *Adoption and Fostering Journal* to say that adoption during the late 1980s in the United Kingdom became "a service for children with special needs" and not for infertile couples.[5] This shortage of babies, which was particularly acute with respect to white healthy neonates, caused infertile parents in the United Kingdom to search additional adoption opportunities elsewhere, mainly in Third World countries, but also in other developed countries, including the United States. And this search for additional adoption opportunities sometimes involves problems, as we will see in the example I bring in this chapter.

Derek Morgan mentions several reasons for the scarcity of babies for adoption in the United Kingdom, by saying:

> The reasons for this decline are commonly linked to the more easy availability of contraception and its more specific tailoring to the needs of individuals concerned; the availability of abortion, and a gradual movement in attitude which has occurred in favor of legal abortion for reasons of preference and a smaller shift in the same direction of reasons of health. Finally changes in attitude to single mother/parenthood and the availability of at least minimal welfare payments make this at least a possibility for women who want to keep their babies rather than to feel they have no alternative but

---

[3] See Radin Margaret Jane. "Market-Inalienability." In: Alpern Kenneth D. (ed.), *The Ethics of Reproductive Technology*. Oxford University Press, New York, 1992, p. 174. This is a short version of Radin's long and rewarding article: Radin Margaret Jane. "Market-Inalienability." In: *Harvard Law Review*, Vol. 100, 1987, pp. 1849–1937.

[4] These details are taken form: Morgan Derek. "Surrogacy: An Introduction Essay." In: Lee Robert and Morgan Derek (eds.), *Birthrights*. Routledge, London, 1989, pp. 73–74.

[5] This quotation is from form: Morgan Derek. Ibid, p. 74. The reference is to the *Guardian*, 12 August 1987, p. 2. Mentioned in Morgan's article in p. 82, endnote no. 67 on.

to make them available for adoption. This is, of course, allied with the role of women in society generally and the view and expectations of them, and them alone as mothers.[6]

This new attitude to single parenthood not only reduced the number of babies that were placed for adoption, but also increased the demand for babies by single women who wanted a child of their own. If these women were not able to conceive, then their preferred choice was to adopt a child. And this trend opened the door to the commercialization of the practice of adoption, even including baby commerce. The benign manifestation of this phenomenon is embodied in "suggestions which propose that pregnant woman who are considering an abortion be counseled or even paid to carry their fetus to term in order to make the newborn child available for adoption by a childless couple."[7] I call this the more benign phenomenon since it could be understood as a commodification of the mother's body (which is itself abominable), and not of the baby. However, this suggestion is problematic from another point of view: the considerations of whether to continue the pregnancy deal with the needs or desires of adults (who want to adopt a baby) and not those of the unborn baby. And if we hold the point of view that the only honorable motive for adoption must be to find appropriate parents for children and not the reverse—not to find appropriate children for potential parents—then this suggested solution to infertility is morally problematic beyond the exploitation of the young woman's body. In any event, there are much more severe manifestations of commercialization of adoption such as the actual buying and selling of babies.

In Chapter 6 which discusses the donation and selling of organs, I presented my approach of strict objection to the selling of organs because of the objectification of the human body that is involved in such commerce. In this case, I deal with an even more radical sort of objectification: not only the trading of organs, but the objectification and commodification of living human beings, though not yet adults. In my previous discussion of Stephen Wilkinson's theory (in Chapter 6), I argued that the most dangerous effect of selling organs is the denial of the intrinsic value of the human being, something which enables the human being to become fungible merchandise. I maintain that there is no difference between adults and children with regard to their fundamental human rights, about which David William Archard claims that they "are too important to be denied or given to some human beings on account of their age."[8] Thus the fundamental rights of children not to be traded, are as valid as those of adults. A major difference is that while adults can object and protest to being sold, children need to be protected by others against such a violation of the above rights. Later in this chapter I will explain why the duty to protect children from being sold should be imposed on the State as their *parens patriae*.

---

[6] Morgan Derek. Ibid, p. 74.
[7] Ibid.
[8] Archard David William. *Children, Family and the State*. Ashgate, Aldershot, 2003, p. 22.

Many people may render this whole discussion about the selling of children as redundant, futile, and unnecessary, and embrace Margaret Jane Radin's viewpoint that "for all but the deepest enthusiast, market rhetoric seem intuitively out of place here, so inappropriate that it is either silly or somehow insulting to the value being discussed."[9] In other words, the banning of this practice is so obvious and apparent that even its discussion abuses the intrinsic value of human beings. However, Thomas H. Murray warns us that "Market enthusiasts claim that the more things we allow the market to distribute, the better off we are. In practice, most market proponents recognize that some things should not be bought and sold. But the moral logic of some market advocates encompasses even children."[10] Thus, we must make crystal clear that human beings including children must stay out of the marketplace, and that human beings are *"market inalienable,"* in Margaret Jane Radin's own words.

In my discussion here, I adopt a sort of Kantian approach toward human beings, of all ages, races, and genders, similar to the attitude ascribed by Stephen Willkinson to Margaret Jane Radin. According to her approach "The person is a subject, a moral agent, autonomous and self governing. An object is a nonperson, not treated as a self-governing moral agent."[11] This approach categorically distinguishes between an object and a subject mainly due to the subject's possession of freedom and rationality. The turning of a person who is a subject, or a free and rational agent into fungible merchandise thus objectifying this person, violates two Kantian Principles. The first is the categorical imperative that requires us to "Act so that you treat humanity, whether in your own person or in that of another, always as an end and never as a means only."[12] This principle prohibits the instrumental use of human beings for other human ends, needs or desires; that is, human beings have intrinsic value as ends in and of themselves, and thus cannot be treated as means. People may not be traded as means, and also not be considered as means for fulfilling someone else's desires or needs. They should be respected as possessing intrinsic value and, therefore, regarded as ends in and of themselves even when they are underage. In Kant's theory, this imperative possesses the status of categorical imperative and thus, has no exceptions or deviations.

The second principle is less familiar than the one mentioned above, but refers more directly to the issue of considering human beings (and more specifically

---

[9] This quotation is from Murray Thomas H. *The Worth of a Child*. University of California Press, Berkeley, 1996, p. 188, endnote no. 6. There he refers to Radin Margaret Jane. "Market Inalienability." In: *Harvard Law Review*, Vol. 100, 1987, p. 1880.

[10] Murray Thomas H. Ibid, p. 35.

[11] This quotation is from Wilkinson Stephen. *Bodies for Sale*. Routledge, London, 2003, pp. 28–29, where he quotes this from: Radin M., "Reflections on Objectification." *Southern California Law Review*, Vol. 65, 1991, p. 345.

[12] This quotation is from Wilkinson Stephen. Ibid, p. 29, where he quotes this from: Kant I. (translated by Beck L.), *Foundations of the Metaphysics of Morals*. Indianapolis, Bobbs-Merrill, 1959, p. 47.

of babies) as fungible merchandise. Kant argues that "In the kingdom of ends everything has either a *price* or a *dignity*. Whatever has a price can be replaced by something else as its equivalent; on the other hand, whatever is above all price, and therefore admits of no equivalent, has a dignity."[13] This explains why there is a categorical distinction between a fungible or interchangeable object—which has or at least can have a nominal price—and a subject—which has a dignity and thus cannot be priced, since in principle it cannot be bought or sold. Therefore, we should not apply market terminology or even market rhetoric to a discussion about subjects, which are characterized by their possession of dignity. According to the Kantian approach, babies also possess the characteristic of dignity and thus cannot be tradable or treated as fungible merchandise. Thus, market terminology and rhetoric are totally inappropriate for any discussion about placing babies or children for adoption, or transferring their custody in any way. Therefore, I do not even need to cite the consequentialist argument that holds that children may not be bought or sold because "the consequences for the children will be bad."[14] I think that the Kantian prohibition is so strict that it overrides any consequential calculations or considerations. This approach considers babies, as well as other human beings, to be what Margaret Jane Radin calls "*market-inalienable*," or something that: "is not to be sold, which is in our economic system means not to be traded in the market."[15] Here I want to stress that not only is the terminology of market rhetoric totally inappropriate, but the fundamental profit principles of the marketplace are incompatible with the fundamental values of the family which are affection, care, love, responsibility, and mutuality. Thus such rhetoric cannot be used even for the discussion of children's care.

However, it should be mentioned that although children are nonsalable, it is legitimate that they be transferred by other avenues such as adoption. But we follow Radin's approach in our insistence that children be immunized/protected against commodification in its broadest sense. Radin says that:

> the term 'commodification' most broadly construed, includes not only actual buying and selling of something (commodification in the narrow sense), but also regarding the thing in terms of market rhetoric, 'the practice of thinking about interactions as if they were sale transactions', and applying market methodology to it. Commodification thus includes owning, pricing, selling and evaluating interactions in terms of monetary cost-benefit analysis or regarding these activities as appropriate.[16]

Radin says that market rhetoric does violence to our conception of human flourishing. She concentrates on the notion of personhood that expresses our commitment to the ideal of individual uniqueness, which contradicts the concept

---

[13] This quotation is from Wilkinson Stephen. Ibid, p. 53.
[14] See Murray Thomas H. Ibid, p. 35
[15] Radin Margaret Jane. Ibid.
[16] Ibid.

that "each person's attributes are fungible, that they have a monetary equivalent, and that they can be traded off against those of other people."[17] Market rhetoric alters our world of concrete persons, who are unique individuals with their own specific personal attributes and characteristics, into a world of disembodied, fungible attributeless entities; a world within which the conception of a person is reduced to an abstract fungible entity with no individual identity or personal characteristics. In Kantian terminology, we can say that market rhetoric negates the person's dignity, and thus, cancels out the person's moral status as an end in and for itself. In the coming example, I present a case of the "twilight zone" that exists between legitimate adoption and illegitimate trade of children.

CASE 1

This case deals with the paid adoption of twins, a case that borders on the buying and selling of babies. The whole story was blown up in Britain when it was publicized that British attorney Alan Kilshaw, 45, and his wife, Judith, 47, purchased twin babies from the American Tina Johnson adoption agency in San Diego, CA, for something like US $12,500. This was in January 2001. Despite the strict prohibition on trading of babies in Britain, the "paid adoption" was possible due to breaches in the adoption laws in Britain and the leniency of adoption laws in the United States of America, particularly in the state of Arkansas.

Another factor that facilitates practices that border on baby commerce is e-commerce on the Internet. Since this medium transcends borders, the Internet facilitates the global adoption of children. It is now relatively easy for people from one country to adopt a baby from a second country, sometimes even through the assistance of an adoption agency from a third country. Such an apparatus serves to override and sometimes violate legal constraints or prohibitions in a certain country by operating the procedure in a place where such prohibitions do not exist (for example, in countries where casinos are illegal, people can gamble through the Internet in virtual casinos whose server is located in countries which allow the operation of casinos). And when this overriding is of adoption laws, it sometimes crosses the line between adoption and baby commerce. This is another example of how technological progress—in this case, e-commerce—outstrips appropriate legislation that is required to avoid distortions in the use of the new technology. Technological progress in many domains, including "online adoption," is more rapid than our ability to suit our moral and practical thinking to the new contingencies it brings about.

However, what makes the above story stunning is that the main agitation among the public as well as the media concerning this story was not about its baby-commerce aspect, but about the behavior of the biological mother, Tranda

---

[17] Ibid, p. 177.

Wecker, 32, from Missouri. Prior to handing the babies over to the British Kilshaws, Wecker gave the same twins for adoption through the same adoption agency to a couple from California, Richard and Vickie Allen, who paid the agency $6000. It was in the 90-day interim period during which the biological mother can change her mind about the adoption (according to Californian law) that the biological mother took the twins, saying that she wanted to say her goodbyes to them. She never returned the babies to the Californian couple. Instead, the agency took the biological mother together with the twins to Arkansas, and contacted the British couple (who had paid in advance for a previous adoption which was not implemented) who traveled to Arkansas and legally adopted the twins there. The agency did not inform the British couple about the previous adoption of these twins; they took the babies home with them to Britain without knowing that they might be involved in unlawful activity. It was only after their return to Britain with the twins that the new adoptive parents discovered that the whole procedure was illegal. The British authorities then assumed supervision for the twins and a long legal procedure ensued—firstly, the cancelation of the adoption by an Arkansas judge, and then a very complicated legal procedure in Missouri—after which the twins were taken to state custody in Missouri. After many additional legal battles, the Missouri Supreme Court decided to give these twins back to their biological mother. This was in March 2004, when the twins were already over 3 years old and had spent most of their life in custodial institutions.

Social workers call the procedure of returning children to their natural families, "restoration." Returning the twins to their biological mother who had originally placed them for adoption was not an easy decision. There were two factors involved: the first was the age of the twins (they were 3 years old already, no longer babies) and their personal history that made adoption very tricky. But the more significant factor was the improvement in the mother's ability to take proper care of her children: her personal situation at that time was much better than when the twins were born. In fact, research has shown that "most . . . children's restorations seemed occasioned by an improvement in family circumstances."[18] In general, it is believed that children are best raised by their natural mothers than to be either readopted or stay in foster families or custodial institutions.

However, I am concerned by the fact that the whole legal procedure concerning the twins' custody was conducted in terms similar to property-rights conflicts, according to which those who paid first are those who have priority on what they paid, that is, the children. They ignored what to me is the pivotal issue: the dangers involved in commercialization of the baby trade, including the "selling" of babies to the higher bidder under the guise of an adoption procedure. In this case, the adoption agency evidently gave the twins to the British couple because they paid more than the American couple, though it must be emphasized that the biological mother did not receive any of the money.

---

[18] Tizard Barbara. Ibid, p. 63.

At the time of the dispute, Sky News exposed the practice of private adoption agencies operating on the Internet and actually offering babies for sale "of all races and colors." Sky News even revealed a price list in which the babies were ranked by price: Chinese babies were "cheapest" while American babies were "most expensive." Even the American babies were "priced" according to their color and their parents' races, the most "expensive" of those were white babies. These Internet agencies even offer "sales" in which babies with mental or physical problems, or babies born to minors or drug addicts, are offered for "extremely cheap prices." Margaret Jane Radin (as quoted by Kenneth D. Alpern) convincingly elucidates the evils of such human commerce as having far-reaching implications on some of our most significant human values, by saying that "if a child can be traded for a Chevy, if such transactions are recognized and allowed in society, then not only [is] the particular child affected, but children in general may be conceived of more in terms of market values—thereby, it is claimed, distorting and limiting their potential for personhood and self-realization."[19] In other words, says Radin, the commodification of human beings (including babies) eradicates their intrinsic value. Alpern says that "if this argument is correct, then even limited commercialization of reproduction in private transactions can be a legitimate concern for social and legal regulations."[20] Thus, social and legal regulations must totally preclude any kind of paid transaction of babies in order to safeguard the humanity and dignity of human entities, including babies, who are market inalienable.

These regulations are needed not only to preclude the actual selling of babies, but also to ban the terms or rhetoric of the marketplace in the discussion of babies. Margaret Jane Radin warns us that:

> When the baby becomes a commodity, all its personal attributes—sex, eye color, predicated IQ, predicated height, and the like—become commodified as well. This is to conceive of potentially all personal attributes in market rhetoric.... Moreover, to conceive of infants in market rhetoric is likewise to conceive of all people they will become in market rhetoric, and to create in those people a commodified self-conception.[21]

We need to relate very seriously to the horrors of the Internet "baby trade" under the guise of adoption, even if this did not play a major role in the current case under discussion. The Sky News story exposed the negative effects of such an immense and worldwide medium on the practice of paid adoption of children. The Internet has made intercountry adoption easier and relatively less expensive

---

[19] Alpern Kenneth D. "Making and Selling Babies: Production and Commerce." In: Alpern Kenneth D. (ed.), *The Ethics of Reproductive Technology*. Oxford University Press, New York, 1992, p. 172.
[20] Ibid.
[21] Radin Margaret Jane. Ibid, p. 186.

than other adoption procedures, enabling adoptive parents to make arrangements through the net before even seeing the adopted children face to face and avoiding the need to travel to the country.

This leads to a discussion of the whole concept of intercountry adoption, which some people maintain is controversial, particularly when these children come from Third World countries. In such cases, those who object to intercountry adoption consider it to be "class, race and national exploitation."[22] There are people who even refer to this procedure as "another form of imperialism."[23] In many cases the procedure is carried out by private agencies, leading to little or no control by the welfare authorities and social workers of both countries. When this is coupled with loose adoption laws (as was our case in the state of Arkansas), then the door to commerce in babies is open. Thus, the juxtaposition of the Internet with the privatization of the adoption process has made the whole world a global market for paid adoption. The hub of paid adoption is profit, and not the welfare, needs or interests of the children involved. In state or governmental adoption authorities, the emphasis is placed where it should be: on the needs of babies and children for adequate care.

It should be mentioned that other parties have more positive views of intercountry adoption. Onora O'Neill, for example, thinks that "The difficulty of adoption in not moreover, as often suggested, a shortage of children for adoption, but rather the legal constraints on cross-border adoption: the world is full of children of all ages who desperately need a loving family."[24] However, she thinks that this should not lower the requirements of competence and eligibility from the adoptive parents, but only to reduce the legal obstacles that are confronted by competent and eligible adoptive parents who attempt to adopt children from other countries. O'Neill stresses that "The legal and regulatory standards that we impose on foster parents and adopting parents do not seem to me to represent an unacceptable restriction on procreative decisions or procreative rights, but rather a reasonable attempt to ensure that the needs of children for at least adequate (and preferably better than adequate) care and nurturing are not put at risk."[25] It is reasonable to assume that there is more, and better, control over the competence and eligibility of adoptive parents in intra-country adoption than in cross-border adoption; in the latter case, the authorities do not have access to information about adopting parents who come from a different country nor can they monitor the care given by these parents to their adopted children after they return to their own country. In private adoption-by-Internet, there is little information given

---

[22] Cole Elizabeth S. "Adoption: History, Policy and Program." In: Laird Joan and Hartman Ann (eds.), *A Handbook of Child Welfare*. The Free Press, New York, 1985, p. 658.

[23] See Ibid. There she brings a quotation from: Benet Mary Kathleen. *The Politics of Adoption*. Free Press, New York, 1976, p. 135.

[24] O'Neill Onora. *Autonomy and Trust in Bioethics*. Cambridge University Press, Cambridge, 2002, pp. 68–69.

[25] Ibid, p. 69.

regarding the adoptive parents except their name and credit card number. And if their main goal is profit and not the welfare of the children, they might not even be interested in additional details.

A close look at the adoption standards of the Child Welfare League of America helps us understand how the shady adoption practices above, distort the original intentions of the practice of adoption. The Child Welfare League considers adoption to be "a mean[s] of finding families for children, not finding children for families. The emphasis is on the child's needs."[26] In other words, the children's welfare and not the parent's needs is the main goal of adoption, since "It is when children are thought of as existing in order to fulfill needs [that] they become, in our minds, commodities."[27] The regarding of children as means to fulfill adults' needs or desires is the first step of their commodification. The only considerations that should have weight in the practice of adoption are those of the children's welfare. Of course, we serve the children's needs best when we take into account the interests and desires of the adoptive parents, but we must match the appropriate parents to the children, and not the reverse—not match the children to the parent's requests. When children are offered to adoption through the Internet, the intended parents remain anonymous to the agency. Since these agencies place children with anyone who pays for them, they are interested in profit and not the welfare of the children involved. This contradicts the central aim of adoption.

It is important to remember that the twins' mother did not receive any payment for giving her daughters to the adoption agency; the commercial relations existed only between the adoption agency and the adoptive parents. This is very significant for our discussion here, since we believe that "Paying individuals for their biological products makes them vendors, not donors. And it places the interaction between the parties squarely in the marketplace,"[28] as Thomas Murray considers the issue of paying the biological parents. Accordingly, the legal situation in the United States is that "paying the mother a fee for adoption is a crime in some states and in others will prevent the adoption from being approved."[29] Paying the biological mother for the adoption would totally change her legal, as well as her moral status, and include her as part of the commercial transaction of her daughters. In our case, the fact that the woman did not receive any payment for the twins was an important factor in the later decision to return the babies to her. It also absolves her from any suspicion of considering them as commodities or fungible merchandise.

---

[26] This quotation is taken from: Cole Elizabeth S. Ibid, p. 641.
[27] Krimmel Herbert T. "Surrogate Mother Arrangements from the Perspective of the Child." In: Alpern Kenneth D. (ed.), *The Ethics of Reproductive Technology*. Oxford University Press, New York, 1992, p. 63.
[28] Murray Thomas H. Ibid, p. 35.
[29] Robertson John A. "Surrogate Mother: Not So Novel After All." In: Alpern Kenneth D. (ed.), *The Ethics of Reproductive Technology*. Oxford University Press, New York, 1992, p. 46.

But there are other cases where the payment-for-adoption situation does exist, even though it may be officially illegal. For example, in Britain and in many other countries "it is illegal for parents to pay for adoption. Despite this, the adopting parents are entitled to make a contribution towards the mother's medical expenses, and these types of payments are increasingly being seen as some kind of 'back-door' payment."[30] And this payment changes the whole moral judgment of the practice adoption. The "back-door" payments appears like a reason to convince the biological mother to place her child for adoption, and thus to insert—even if through the back-door—the market values and rhetoric into the transaction of babies. And this should be strictly rejected. A similar problem is that of the private adoption agencies who are paid by the adoptive parents. They cannot claim that the money they receive is to cover expenses, since they are commercial agencies whose principle is maximizing profit for their owners. And this profit comes from the difference between the money they receive from the adoptive parents and their expenses (since the biological parents do not receive any part of this payment). This profit is called in Marxist terminology "surplus value," and is created by people's labor. Since this profit is gained from trade in children, the entire transaction becomes commercial and the children themselves become analogous to merchandise.

The appalling point of the case above is not that fraud was involved and thus, the twins were sold twice to two different purchasers, but instead—that they were sold to begin with and regarded as objects for trade or commercial activity. This should not be regarded as a case of simple fraud, as happens frequently with the selling of cars, houses, real estate, or other kinds of property. We condemn fraud as a violation of the fundamental norms of free trade—but the very terminology of fraud does not apply to babies as the very concept of human sale or trade is abominable and monstrous.

I already mentioned in Chapter 6 of this book that we usually ascribe at least two values to a product: its use value and its exchange value (which is its market or trade value). The use value of an umbrella, for example, is protecting its user against the rain. Its exchange value can be one scarf, or $5. Obviously, the exchange value of the same umbrella might fluctuate under different circumstances since prices are determined according to the rules of supply and demand. Therefore, a surplus of umbrellas in a certain geographic area might, for example, bring down the price.

But human beings of all races, genders, colors, and ages, possess intrinsic value. This intrinsic value should not be determined by their social status nor even by their work (although people's work has exchange value from the employer's point of view, this is not the sum value of their worth as human beings). We ascribe intrinsic value to people simply because they are human beings, and ascribe equal

---

[30] http://www.netfreedom.org/news.asp?item=139.

value to all mankind. Once we determine the value of human beings according to market conditions (as was the case in slavery), we commodify them and brutally strike at their very humanity. This, in turn, harms mankind as a whole. And this is the same thing as pricing children for adoption according to their race, color, and social background of their parents. A decent person cannot remain indifferent to this immoral phenomenon.

The practice of adoption is meant to provide decent and appropriate parenting and welfare to children whose parents do not or cannot provide what they need, thus recognizing that adopted children possess the same intrinsic value as do other human beings. Society acknowledges this value of children by establishing institutions, such as adoption authorities, which entrust these children to the hands of appropriate parents. And these authorities traditionally used to be governmental institutions who had no intentions of making profits and were completely obligated to the children's humanity and best interests.

The authorizing of private bodies to deal with the adoption of children—and particularly of babies—has opened the door to the specter of agencies who ignore or subsume the needs of the children to the profit-motive by placing children with the highest bidders. The needs of children are subordinate to "market rules," thus raising doubts regarding the morality of paid adoption and of private adoption agencies.

The only way to ensure that children will be considered as possessing intrinsic value, as ends in and of themselves, is to be more strict with the prohibition of the trading of human beings. Each country's authorities must be required to apply and enforce this ban on any possible medium and any form of trade. In order to avoid the temptation of immense profit to be made by private agencies, these type of agencies should be outlawed. Instead, adoption-procedure arrangements should be entrusted exclusively to governmental authorities who try to ensure basic human rights of every resident within their territory. In the case of children, the governmental authorities have the responsibility of protecting children's fundamental rights as a result of the state's status as the children *Parens Patriae*, which was explained in previous chapters. It is not enough that all countries should legislate laws that ban the trading of human beings (something which already exists around the world), but that adoption should be banned by external bodies or private agencies that are not the state welfare authorities. Although such legislation may not totally eliminate commerce in children it would significantly reduce its scope to illegitimate criminal groups. The fact that private adoption agencies today are legally permitted to exist and regulated, endows them with a sort of legitimacy and dulls the public's social condemnation and moral disapproval.

According to Kenneth D. Alpern's interpretation of Margaret Jane Radin's position, "proper conception of human personhood and human flourishing (what is to live a good and rewarding life) cannot be secured if certain aspects of life are regarded as subjects of economic activities and values."³¹ The fact that babies are adopted through profit-making private agencies violates the "human personhood and human flourishing" not only of the children involved in such transactions but also of humanity as a whole. Alpern explains that in Radin's view "not only do economic *activities* (being monetarily priced, bought and sold, etc.) have a role in determining what a thing is and can do, but so too do the ways these things are *thought* and *spoken* about. If this is true, then it can be argued that the effects of economic activities and values extend beyond the few individuals actually involved in the economic transaction."³² This activity harms our comprehensive conception of a person who has dignity, and thus cannot be measured or even discussed in market values and market rhetoric. David William Archard says "Being human does matter and it is precisely because they are human beings, albeit young ones, that children are entitled to be treated in ways that nonhuman, such as animals, are not."³³

We already saw in previous chapters that the state functions as the children's *Parens Patriae*, and thus, should protect their interests and "in its child protection (CP) practices, show a concern for the child's best interests."³⁴ Archard stresses that "The state has a duty as *parens patriae* to safeguard the interests of children when they are unprotected."³⁵ However, states have obligations to all children, not only toward the children within their jurisdiction. David William Archard expresses this view by saying that:

> The United Nations Convention on the Right of the Child is an international charter, and those national governments that are its signatories have responsibilities towards *all* children of the world. States exist within an increasingly interdependent world bound together by multiple economic, political and cultural bounds. States cannot be indifferent to the global effects, both direct and indirect, of their own actions and of those citizens, organizations and companies within their jurisdiction.³⁶

The example discussed in this chapter regarding intercountry adoption (and even more clearly, the example of Elian Gonsales in Chapter 8 of this book) demonstrate the pressing need for international cooperation in protecting children's rights and avoiding commerce in children.

Every country is enjoined to establish a CP practice which consists of "a set of laws and policies enforced and implemented by a wide variety of official and semi official agencies.... The agencies responsible for the implementation

---

[31] Alpern Kenneth D. Ibid, p. 171.
[32] Ibid, p. 172.
[33] Archard David William. Ibid.
[34] Ibid, p. 117.
[35] Ibid, p. 126.
[36] Ibid, pp. 118–119.

of CP practice include social work, educational, police, legal and correctional, and medical bodies, all acting frequently in structured modes of interagency cooperation to protect children."[37] The agencies that implement CP practice, typically protect children from abusive adults such as the children's parents and guardians. I maintain that another role of these agencies should be to arrange the adoption of children.

---

[37] Ibid, p. 142.

SECTION D. PUNISHMENT

Several decades of theoretical discussion concerning the concept and social function of punishment within society have yielded a number of important principles, leading to a reasonable understanding of punishment. These principles are explicated in this section, and include: the proportionality principle (in which the punishment must fit the crime), elements of deterrence (as in the consequential or utilitarian approaches), and humane treatment of convicts. However, several marginal and problematic situations still remain where additional considerations must be entertained, situations that vex the ethicist bent on making theory adhere to praxis. One acute problem in the current practice of punishment is that it is imposed by imperfect legal systems, whose obligation is first and foremost toward the rule of law, and not necessarily to moral or ethical considerations. This may cause distortions in the implementation of punishment, such as overly lenient penalties (in domestic abuse, for example), and exceedingly severe ones (such as the death penalty for the mentally retarded). I examine the complexity of these issues in the different cases I bring up in this section.

CHAPTER 11: PUNISHMENT OF SEX OFFENDERS

This chapter first discusses general characteristics of punishment as a whole as the background for the expectations we have for the punishment of sex offenders. These characteristics include the proportionality principle in which the punishment must fit the crime However, I take into consideration that our usual guidelines regarding punishment are somewhat insufficient when we discuss punishment for sex crimes, due to the propensity of these types of crimes to arouse extreme emotional reactions.

I discuss two aspects of punishing sex offenders. The first deals with the outrageously mild punishments that are meted out to sex criminals in Israel, both in length of incarceration and in severity of punishment (sometimes minimized to community service), particularly to teenage sex offenders. This strikes at our deepest instincts regarding the proportionality principle in which the punishment must fit the crime. It also begs the question whether the tacit acceptance of society of these mild punishments, indicates that society accepts these values and norms behind the punishments involved. Do most people agree that men are superior to women, and that violence is a legitimate means for achieving men's goals and desires? If society itself does condemn such offences, then it follows that the courts that represent this society, should impose much more severe penalties for sex crimes.

The second aspect deals with the offering of plea bargains by the prosecution, in which curative treatment to the sex offender is a stipulation for reduced punishment. It is a relatively new idea of mitigated punishment pending the criminal's consent to chemical castration. In this case I argue that rehabilitation is not part of the prosecution's role as a legal authority.

CHAPTER 12: PUNISHMENT AND DOMESTIC VIOLENCE

Domestic violence is not just a sub-category of violence in general. It is inherently complex because there is an unclear border or thin line between punishment (disciplining children) and actual violence (beating children). Another problem is that when the offender is a parent, for example, then punishing him or her is likely to harm the family at large, such as by taking away their means of support. These complex issues are often used by the authorities as an excuse to abstain from pursuing and punishing offenders. Consequently, they renege on their two obligations—to punish criminals and protect the weak. I present the argument in this chapter that the privileges of autonomy and non-interference generally accorded to the family by the State, are privileges that are conditional on the proper functioning of the family. When domestic violence rears its head, these privileges of autonomy are cancelled and the family is, indeed, subject to State control and interference.

The second problem, more specific to Israel, is the intolerable laxity with which domestic violence is treated by institutions in Israel, most notably by the legal and law enforcement systems. In general, the legal system in Israel exhibits outrageous clemency toward criminals in general, and for domestic violence in specific.

CHAPTER 13: CAPITAL PUNISHMENT AND THE MENTALLY RETARDED

In this chapter I focus not on the general debate regarding capital punishment, but on applying the death sentence to mentally retarded or mentally ill criminals. One of the most relevant terms for discussing the legal and moral accountability of the mentally retarded is that of *mens rea* (intent required to commit the crime). I argue that this notion of diminished responsibility should preclude the imposition of capital punishment on mentally retarded and mentally ill people, without entering into the ideological dispute regarding capital punishment per se.

Examples are given to illustrate the troublesome phenomenon of Texas courts that are not much swayed by the impaired mental condition of offenders when deciding to impose capital punishment on them. I conclude that even those who support capital punishment should demand that it be imposed not only on those who deserve the most severe penalty allowed by society, but also those with the highest level of *mens rea*. Thus when trying mentally retarded or insane criminals, their mental state should be considered as a mitigating factor in reducing the death penalty to life imprisonment.

Chapter 11

PUNISHMENT OF SEX OFFENDERS

Before turning to discuss the subject of this chapter—punishment of sex offenders—it is important to discuss the general characteristics of punishment as a whole. Antony Flew holds that punishment must contain at least five elements that distinguish this term from other hardships people may suffer from. First, the person who is punished must consider it a hardship or unpleasantness; this means that a punishment cannot be something beneficial to the punished person (I will argue later in this chapter that the rehabilitation of offenders cannot be considered punishments). Second, it should be a reaction to an offense, meaning that people should not be arbitrarily punished for no reason. Third, it should be inflicted only on the guilty offender and not imposed on innocent victims or suspects whose guilt has not been proven. This is a very important component, since it prohibits us from arbitrarily punishing persons who did not commit a crime or an offense. Fourth, punishments do not include natural disasters that harm human beings, but are administered by human beings. Fifth, punishments must be imposed by virtue of special authorities conferred by a system of rules, against which the offense has been committed; a person cannot take the law into his own hands to beat up a cheating neighbor, for example.[1] This general concept of punishment provides the background for certain expectations we have for the punishment of sex offenders.

Our usual guidelines and principles regarding punishment are somewhat insufficient when we discuss punishment for sex crimes, due to the propensity of these types of crimes to arouse extreme emotional reactions. In addition, there is often an assumption that sex offenders are not "regular" criminals but are "diseased,"

---

[1] These elements of punishment are In: Flew Antony. "The Justification of Punishment." In: *Philosophy*, Vol. 29, No. 3, October 1954, pp. 293–295.

"ill," or "sick." This chapter, hence, deals with two related issues in which these points surface. One is the tendency to reduce the sentences of teenage sex offenders, both in length of incarceration and in severity of punishment (sometimes minimized to community service). The other is the relatively new idea of mitigated punishment pending the criminal's consent to chemical castration. Both are complex in that not only do they expose ambivalent positions vis à vis sex crimes, but they problematize the concept of punishment itself.

In this chapter, I discuss two aspects of punishing sex criminals. The first deals with the outrageously mild punishments that are meted out to brutal sex criminals, something which strikes at our deepest instincts regarding the proportionality principle in which the punishment must fit the crime. The second example deals with the offering of plea bargains by the prosecution, in which curative treatment to the sex offender is a stipulation for reduced punishment. In this case, I argue that rehabilitation is not part of the prosecution's role as a legal authority.

CASE 1

This case deals with the phenomenon of imposing extremely mild penalties on sex offenders in Israel and is at least partly relevant to other countries where penalties are also relatively mild. The first case deals with a group of teenagers who brutally raped a high school girl over a period of several months and were sentenced in the Tel Aviv District Court to very short periods in prison. The two leading offenders were sentenced to 4 years in prison, another one was sent to prison for two and a half years, and most outrageous of all was that the other four convicted offenders were not even sentenced to imprisonment, but only to 6 months of community service. These were the punishments for committing numerous gang rapes of a young girl. There are many other examples of scandalously mild punishments for sex offenders, such as the 6 months of community service and 8-month suspended prison sentence imposed (by an Israeli Natanya court) on a 24-year-old man convicted in February 2005 for nine cases of sexual attack of girls between the ages of 6–12 years old. The offender did not spend even 1 day in prison for the attacks, and it is almost anti-climactic to say that during the period of community service, the man sexually attacked a 7-year-old girl twice. Another horrific example is the case of a truck driver who tried to rape a young female hitchhiker. The young woman managed to escape her attacker by jumping out of the truck, but unfortunately she was run over by another truck and a car and was killed. In June 2005, the attempted rapist was sentenced by the Court of Hadera to only 1 year in prison (of which he will be incarcerated for no more than 8 months). These three representative examples point to a widespread phenomenon regarding the social norms and values of a State which metes such mild penalties for sex crimes.

We must view mild punishments within the context of a general discussion regarding the roles and functions of punishment in society, and the implications of imposing of very mild punishments for certain crimes. I divide the theoretical discussion of punishment into two: the justification of the institution of punishment itself (or what John Rawls calls "the practice"[2]), and the justification of a particular punishment within this practice.

There are three broad categories of justification for the practice of punishment as a whole. The first is consequential or utilitarian; the second deals with considerations of justice; and the third involves considerations of social values, norms, and virtues. The utilitarian justification is a forward-looking approach that justifies the practice of punishment in terms of the benefit or utility to be gained from imposing penalties on offenders. The most obvious advantage is that excluding an offender from society by incarceration, prevents criminals from committing additional crimes during the period of their prison sentence. The other benefit of punishment is that of deterrence. A criminal who spends time in jail will try to avoid committing acts in the future that will put him back behind bars. Punishment also deters other members of society if they become convinced that crime does not pay and that if they break the law they are likely to be caught and sentenced. The assumption behind the forward-looking approach is that all human beings, even hardened criminals, make cost-benefit calculations and assess their risk of getting caught before committing a crime. If the punishment is extremely unpleasant, and the risk of getting caught is high, people in general are much less likely to risk breaking the law.

The second justification of the practice of punishment is retribution or repayment. This is a backward-looking approach wherein punishments are justified on the grounds that the criminal has created an imbalance in the social order that must be addressed by action against the criminal. When a person causes evil, society should make the person "pay" for his acts by an equally unpleasant punishment; thus the proponents of this approach believe that punishment is the intrinsically appropriate response for past wrongdoing.[3] Another version of this justification argues that the harm which is caused to society by the offender, is balanced by a proportional harm which is caused by society to the offender in return. Thus according to this approach, punishment is a kind of appropriate retaliation that suits the act for which the penalty is imposed. However, the gravity of the offense and the severity of the punishment must be proportional and appropriate: penalties should be mild for slight offenses, and severe for cruel and malicious ones.

The central basis of any retributive theory is that of guilt. According to Jeffrie G. Murphy, "Retributive theories of punishment maintain that criminal guilt merits

---

[2] See Rawls John. "Two Concepts of Rules." In: Baird Robert M. and Rosenbaum Stuart E. (eds), *Philosophy of Punishment*, Prometheus Books, Buffalo, New York, 1988, p. 37.
[3] See, for example, Cottingham John. "Punishment." In: Becker Lawrence C. and Becker Charlotte B. (eds). *Encyclopedia of Ethics*. St. James Press, London, 1992, pp. 1053.

or deserves punishment, regardless of considerations of social utility."[4] One of the most extreme versions of the retributive approach is exemplified by Kant, who considers guilt as *sufficient* condition for justifying punishment. Murphy quotes the following from Kant:

> "Even if a civil society were to dissolve itself by common agreement of all its members (for example, if the people inhabiting an island decided to separate and disperse themselves around the world), the last murderer remaining in prison must be executed, so that everyone will duly receive what his actions are worth and so that the bloodguilt thereof will not be fixed on the people because they failed to insist on carrying out the punishment; for if they fail to do so, they may be regarded as accomplices in the public violation of legal justice."[5]

Kant holds that society is duty-bound to punish criminals in order to restore justice. Thus when a society refrains from punishing wrongdoers, or imposes only a mild punishment for an abominable crime, then that society is guilty of complicity for the crime committed by the perpetrator. Often, retributive theory is thought to be linked to *lex talionis*. Literally "law as retaliation" in the Latin language, lex talionis is the belief that the purpose of the law is to provide retaliation for an offended party ("an eye for an eye"). Wesley Cragg reminds us that judges often refer to this principle when handing down their sentencing: "It is not uncommon for judges to describe the sentence as demonstrating the court's view of the seriousness of an offence."[6]

The third justification is neither forward nor backward looking but merely assumes that society uses punishment to express its loathing or repugnance of the act or crime for which the penalty is imposed. David Lewis refers to such an approach when he says that "The guilty ought to suffer a loss... as an expression of our abhorrence of their offences."[7] The severity of the punishment expresses society's norms and attitudes toward actions or crimes of the same kind. The violation of a trivial norm in society receives a mild punishment, while violation of an important, crucial norm receives a severe punishment. This is similar to the proportionality principle of the retributive approach, with one difference: here, the severity of each offense is determined within the norms of the specific society. Norval Morris and Donald Buckle express this view by saying that: "The criminal Courts have traditionally represented the common man and the common man's view of morality. The Judges have earned the confidence of the

---

[4] Murphy Jeffrie G. *Retribution, Justice and Therapy.* D. Reidel Publishing Company, Dordrecht, 1979, p. 77.
[5] This quotation is from Murphy Jeffrie G. ibid, p 82.
[6] Cragg Wesley. *The Practice of Punishment.* Routledge, London, 1992, p. 169.
[7] Lewis David. "Do We Believe in Penal Substitution?" In: *Philosophical Papers*, Vol. XXXVI, No. 3, November 1997, p. 204.

people as unbiased and incorruptible men."[8] The idea that the courts represent the common man's view of morality, dictates that judges' decisions should reflect the common morality and fulfill the common man's expectations for a certain level of punishment—even when a judge may personally hold a different view in his own heart or conscience.

Andrew von Hirsch argues that "if punishment conveys blame, it would seem logical that the quantum of punishment should bear a reasonable relation to the degree of blameworthiness of the criminal conduct."[9] Antony Duff and David elucidate his view and say, "Von Hirsch founds the principle of proportionality on an account of punishment as censure. Punishment should express the censure which criminals deserve; and if it is to express the appropriate degree of censure, its severity must be proportionate to the seriousness of their offences."[10] This means that the censure of an offence is derived from the social assessment of its gravity, and this should determine the extent to which the offender should be punished. If a society considers a certain offence as only mildly blameworthy or minor, the punishment for it should be mild.

I would like to examine the flagrantly mild sentences for sex offences within the context of each of the three justifications for the practice of punishment, as brought above. If we view it through the prism of utilitarianism, it is clear that a community service penalty would not deter anybody from committing any crime, certainly not cruel crimes such as gang rape or other kinds of sexual attack. Thus the deterrent principle falls by the wayside, leading us to believe that it was not taken into account by the judges when they made their decision. Mild punishments for severe crimes certainly do not transmit the message that "crime does not pay."

Presiding Judge Nathan Amit, in the case above, referred in his verdict and sentencing to the importance of "proportionality between the gravity of the offence and the severity of the punishment," and then called the case an example of "the prohibited, the criminal, the venal, the bestial." Why, then, were the punishments so light, when the judges themselves stated that the rapists "looted [the victim's] purity, her joie de vivre, her hopes for the future, her feeling of self respect, and self esteem"? It must be clarified that sentences for jail-time handed down by judges in Israel for violent crimes (including murder) are, and have long been, significantly lower than what is accepted in the United States and most of the democratic world. Therefore sex crimes fall into this category, and judges seem to toe the line in this regard. The following Chapter 12 will discuss this in more depth.

---

[8] Morris Norval and Buckle Donald. "The Humanitarian Theory of Punishment: A Reply to C. S. Lewis." In: Grupp Stanley E. *Theories of Punishment*. University of Indiana Press, Bloomington, 1971, p. 312.
[9] Von Hirsch Andrew. "Censure and Proportionality." In: Duff Antony R. and Garland David. (eds.), *A Reader on Punishment*, Oxford University Press, New York, 1994, p. 118.
[10] Duff Antony R. ibid, p. 112.

Ten tells us that "*lex talionis* attempts to provide a simple basis for arriving at a specific punishment for each crime, and the principle embodied in it is that the punishment should inflict on the offender what he has done to his victim."[11] It seems clear that the judges themselves did not operate according to the *lex talionis* principle, since they imposed mild punishments on a crime that they themselves considered bestial and brutish. And they could not have meted the punishment according to the retribution theory either, which "is embodied in Hegelian claim that punishment 'annuls' the crime,"[12] because the crime here is severe and not proportional to the light sentence. Thus we have to conclude that the judges did not use any retributive considerations at all.

The only remaining alternative is to relate the judges' decision to the third justification for punishment that deals with social values, norms, and virtues. It seems to me that when a society becomes accustomed to violence, that same society, including its judges, also becomes inured to the suffering of victims of that violence especially when they are women or belong to a minority group. It should be mentioned here that in a previous case of gang rape, the Northern District General attorney and the Prosecution of the Northern District initially decided not to submit indictments against the rapists. The official reason for that decision was "lack of public concern," and only when women's organizations put pressure on the Israeli Supreme Court did the prosecution finally decide to indict the rapists. We have to ask whether society accepts these values and norms, and thus tacitly accepts the mild punishments meted out to these offenders. Do most people agree that men are superior to women, and that violence is a legitimate means for achieving men's goals and desires? If society itself does condemn such offences, then it follows that the courts that represent this society, should impose much more severe penalties for sex crimes.

Another goal of punishment is "educating the public." Nigel Walker cites the "educative effect" of punishment as a useful outcome which serves to justify punishment as a whole.[13] Jean Hampton goes even further and suggests that "by reflecting on the educational character of punishment we can provide a full and complete justification for it."[14] Even if the educative function of punishment is not used by ethicists as a frequent justification for this practice, Wesley Cragg says that "sanctions are often justified in common parlance for their educative value, a theme as old as punishment itself."[15] If we examine the possible educative message of such punishment or the values and norms which they should promote, we reach the horrifying conclusion that the majority of males do not think that

[11] Ten C. I. *Crime, Guilt and Punishment*. Clarendon Press, Oxford, 1987, p. 151.
[12] ibid, p. 38.
[13] Walker Nigel. *Why Punishment*. Oxford University Press, Oxford, p. 21.
[14] Hampton Jean. "The Moral Education Theory of Punishment." In Simmons John A, Cohen Marshall, Cohen Joshua, and Beitz Charles R. (eds.), *Punishment, A Philosophy & Public Affairs Reader*, Princeton University Press, Princeton, 1995, p. 113.
[15] Cragg Wesley. ibid, p. 169–170.

gang rape deserves social denunciation,[16] and therefore, the legal system should follow this normative standard, social values, and social moral atmosphere.

However, there is also a large public that is disgusted by these crimes, especially when they are committed against the weak and the defenseless/underprivileged members of society. And this group has to make every effort to alter prevailing social norms in order to protect and preserve "the dignity, purity, joie de vivre, hopes for the future, feeling of self respect and self esteem of young girls" (as well as other disadvantaged sectors in society)—expressions cited by the very judges who handed down the absurdly light punishments of the current case. Thus, this public should make every effort to change the standards of punishment for sex crimes and demand the enforcement of the fairly new law, which requires imposing at least a certain amount of prison time on sex offenders—half of the maximal period allowed by the law (a law which the Israeli judges simply ignore when they mete out the penalty of community service). In my opinion, this minority should also agitate for longer periods of incarceration for perpetrators of sex crimes. Changing of the levels of punishment might change public norms and attitudes towards these heinous crimes; at the very least, it would prevent laughable penalties such as community service to be meted out for gang rape. Even if the majority would cry that such a change in legislation would not reflect "the people's will—I respond that sometimes 'the people' and 'the majority' might be wrong."

The demand that feelings of outrage and disgust (or in the words of Christopher Harding and Richard W. Ireland, that punishment "may reflect a mood of anger or frustration"[17])—should be expressed in the verdict and punishment, is not only a demand to re-establish justice. It expresses the idea that ignoring public anger and frustration about mild punishment for abominable crimes, is not merely social insensitiveness or obtuseness, but a clear sign of social bestializing. When punishment does not fulfill its function as a social practice, people lose confidence not only in the legal and judicial institutions but also in the morality of society as a whole. According to Wesley Cragg and others,

> "Social protection of protecting the public has become the dominant view of the purpose of sentencing in recent years."[18] When this function of protecting society (including each individual within it) and protecting the public order is not sufficiently fulfilled, those who live in that society, remain unprotected and vulnerable. They lose faith in the enforcement of law and in the unstable and disintegrating social order,

---

[16] When Nigel Walker discusses Feinberg's "Expressive Function of Punishment" in Feinberg's book *Doing and Deserving* (Princeton University Press, Princeton, New Jersey, 1970), he ascribes Feinberg the idea that denunciation is the justification for punishment, even though not the ultimate justification. See Walker Nigel. ibid, pp. 23–24.

[17] Harding Christopher and Ireland Richard W. *Punishment: Rhetoric Rule and Practice*. Routledge, London, 1989, p. 107.

[18] Cragg Wesley. ibid, p. 176.

and are more easily drawn into violence. If we do not want Jerusalem to turn into Sodom, we must changing the standards of punishment for violent crimes, including sex crimes.

The following example turns to a slightly different issue the role and involvement of the prosecution in rehabilitative forms of punishing sex criminals.

CASE 2

This case deals with a plea bargain offered by the Israeli prosecution to a perpetrator who was convicted of committing a rape. Usually, plea bargains are offered to suspects before and during the trial, and most of them offer reduced punishments to a suspect (after the conviction) in return for a confession.

However, in our current case the prosecution's offer to the convict included a few additional (and very unusual) conditions. It stipulated that the convict would agree to receive psychological and medicinal treatment for suppressing his sexual impulses during his incarceration, and would keep on receiving the same treatment for a period of 3 years after his release. If he chose to refuse this treatment, he would be denied a parole board and his punishment would not be reduced by a third (all prisoners are entitled to this reduction in prison time for good behavior). In other words: he would receive a relatively severe penalty in the first place, and this would not be shortened by a parole board during the incarceration.

The concept of linking the duration of imprisonment and a curative treatment during incarceration is not unknown in modern criminology and sociology. At its basis is the hypothesis that criminal behavior results from the maladjustment of offenders to their vicinity or society. Thus imprisonment is not only punishment but should be aimed at rehabilitating criminals and preparing them to re-enter society, with its social norms and requirements, after incarceration. This rehabilitative approach encompasses the "reform" as well as "curative" viewpoints, which are similar but not identical. The reform approach attempts to reform the character of the offenders, or at least their behavior, while the curative approach is even more far-reaching and ambitious in its attempt to cure the offenders. Thus the rehabilitative approach wants to utterly modify the whole system of punishment by presuming that criminal behavior is only a symptom of a psychological disorder that needs to be treated and cured, and that traditional punishments should be substituted by therapeutic or medical models of treatment.[19]

Both the reform and curative approaches are motivated by the desire to reduce the harm caused to society by criminal behavior, as well as the societal harm caused by the punishment meted out to the criminals afterwards. Those who

---

[19] This difference between the two similar approaches is taken from: Cottingham John. ibid, pp. 1054.

support these approaches "are not so much interested in 'punishing' the offenders (which is causing additional harm to the offender), as in avoiding future harm that might be caused by these and by other offenders. Hence, they are much more concerned with the underlying reasons and causes of the crime than in the retributive aspects of punishment. Accordingly, they concentrate essentially on the social and psychological, rather than the moral and ethical aspects of the offender."[20] From their point of view, the moral issue of guilt is secondary or even irrelevant to the decision about what should be done with the convicted offender. They are not content with punishment alone, but instead aim to treat or cure criminal behavior and its underlying causes.

Stanley E. Grupp summarizes the rehabilitative approach by saying:

"the keynote of the approach is, of course, the individualization of punishment—perhaps we should say treatment—and working with the individual in such a way that he will be able to make a satisfactory adjustment, or at least non-criminal adjustment, once he is released from the authority of the state. Since most offenders do return to society, and some never technically leave it, it makes good sense to work with the offender in such a way that he will not again be a criminal liability."[21]

The fundamental principle behind any rehabilitative approach is the imposing of a flexible time schedule for incarceration, usually with minimum and maximum time-limits such as "between 2 and 5 years in prison."[22] During this period of time, prisoners are supposed to progress through a rehabilitation process that provides them with better tools for integration into the time-abiding social framework. This treatment generally includes psychological intervention as well as providing professional skills and normative habits to enable the prisoner to re-enter society. Parole boards are supposed to determine the date of release of such prisoners from prison according to the progress of their rehabilitation treatment. Note: A large-scale rehabilitation approach was undertaken by the United States many years ago and largely discredited by the 1970s; it proved to be extremely expensive and did not make a significant impact on "reforming" or "curing" the criminals. However, certain elements of the approach have survived to this day.

The problem with the rehabilitative approach is that it effectively negates the concept of freedom of choice. It assumes that criminals act as they do because of some kind of maladaption in society, and not because of moral decisions or conscious choices. In other words, an offender is not a criminal but a sick person who needs to be healed. Once we presume that offenders have no control over

---

[20] Ezra Ovadia. *The Withdrawal of Rights*. Kluwer Academic Publishers, Dordrecht, 2002, p. 131.
[21] Grupp Stanley E. ibid, p. 8.
[22] On more details on this principle, and on the rehabilitation approach as a whole, see Rachels James. (ed.), *Moral Problems. A Collection of Philosophical Essays*, 3rd edn. Harper & Row, New York, 1979, pp. 315–317.

their behavior and commit crimes due to social circumstances or psychological reasons, then they have no free choice and should not be punished. Therefore, we should not punish them or assign them any moral guilt, but commit them to special rehabilitation centers.

The rehabilitation approach is sometimes called "The Humanitarian Theory of Punishment" and its foundations rest on the shaky basis of the elimination of moral guilt. C.S. Lewis argues: "The Humanitarian theory removes from Punishment the concept of Desert. But the concept of Desert is the only connecting link between punishment and justice."[23] The removal of guilt from the offender, involves an acute moral difficulty since "this doctrine merciful though it appears, really mean that each one of us, from the moment he breaks the law, is deprived of the rights of a human being."[24] The assumption that people commit crimes because they are sick or unable to make choices, reduces their moral standing and eradicate their moral responsibility. This, in turn, degrades their moral status to something less than a moral agent—a very problematic issue. Lewis says that "when we consider what the criminal deserves and consider only what will cure him... we have tacitly removed him from the sphere of justice altogether; instead of a person, a subject of rights, we now have a mere object, a patient, a 'case'."[25]

Herbert Morris presents this moral difficulty in a more blatant wording:

> "When we treat an illness we normally treat a condition that the person is not responsible for. He is 'suffering' from some disease and we treat the condition, relieving the person of something preventing his normal functioning. When we begin treating persons for actions that have been chosen, we do not lift from the person something that is interfering with his normal functioning but we change the person so that he functions in a way regarded as normal by the current therapeutic community. We have to change him and his judgments of value. In doing this we display a lack of respect for the moral status of individuals, that is, a lack of respect for the reasoning and choices of individuals. They are but animals who must be conditioned."[26]

This analysis of the rationale behind the rehabilitation approach brings Morris to conclude that "The primary reason for preferring the system of punishment as against the system of therapy might have been expressed in terms of the one system treating one as a person and the other not."[27] For Morris the moral problem with such an attitude towards an offender is the denial of significant human characteristics of that offender. In any event, this is not the reason that I oppose the offer cited above made by the prosecution to the offender.

---

[23] Lewis C. S. "The Humanitarian Theory of Punishment." In: Grupp Stanley E. ibid, p. 302.
[24] ibid.
[25] ibid.
[26] Morris Herbert. *On Guilt and Innocence*. University of California press, Berkeley, 1976, p. 43.
[27] ibid, p. 47.

I now return to the current case: the rather bizarre combination of plea bargain and rehabilitation program offered by the prosecution to the offender. I understand why it would appeal to the offender, but I am puzzled why it would appeal to the prosecution, since such an offer seems to take the curative approach that eradicates the offender's moral guilt.

But in addition to this (meta?) ethical problem is the professional problem regarding the social role and function of the prosecution, to which I will return to later in this chapter. Lewis tells us that "The Humanitarian theory, then, removes sentences from the hands of jurists whom the public conscience is entitled to criticize and places them in the hands of technical experts whose special science do not even employ such categories as rights or justice... this transference results from an abandonment of the old idea of punishment, and therefore of all vindictive motives."[28] The prosecution, as part of "the old idea of punishment," cannot cooperate with this tendency, since it is not a professional authority in any therapeutic discipline. The prosecution belongs to the judicial part of society, and thus, the part which represent the public interest in making justice, and it should not represent interests of others, such as the defendant, the welfare or the social authorities. These authorities are those who might raise therapeutic offers to the prosecution and not the contrary.

The support made by the prosecution of such an idea undermines the principal ground upon which it bases its prosecutions. If there is no moral guilt, then there are no grounds for accusing people of committing crimes and definitely no grounds for demanding their punishment. At most, there are grounds for demanding curative treatment. But the essence of prosecution lies in prosecuting offenders, convicting criminals, and demanding their punishment. The rehabilitative approach is simply not consistent with the prosecution's mandate. The eradication of guilt from offenders is not consistent with the prosecution's social role and social function. Once moral guilt no longer exists, there is no reason to punish the criminal and no space for the prosecution's function.

To clarify the inappropriateness of the support of the prosecution in the rehabilitation process I bring Stanley E. Grupp's explanation of the discrepancy between the idea of punishment and the idea of rehabilitation. He says:

> "Some Persons taking the rehabilitative stance assert that they are "anti-punishment;" they speak of the 'crime of punishment' and further contend that the rehabilitative objective is not punishment. Consistent with the definition of punishment as the objective of the state's handling of the adjudicated offender, the 'anti-punishment' point of view *is* appropriately considered under the rehabilitative rubric."[29]

---

[28] Lewis C. S. ibid, p. 303.
[29] Grupp Stanley E. ibid, p. 9.

When the prosecution takes an "anti-punishment" stance it undermines its jurisdictional mission as the body in charge of suing the offender. This is not consistent with its social role and function.

An additional problem with this offer regards the linkage between the shortening of the imprisonment and the treatment which the offender is supposed to receive. If a certain punishment appears in the prosecution's eyes as appropriate to the severity of the offence, its shortening is reasonable if we substitute the remainder period of the imprisonment by another penalty. This serves to preserve the principle of balance between the crime and the punishment. What we might understand from the prosecution's attitude is that the curative treatment is a substitute for the shortening of the imprisonment. This attitude is no less bizarre than the approach which removes guilt from the offender. The rendering of a medical or curative treatment (both medicinal and psychological) as a sort of punishment (or substitute for punishment), is inconsistent with our usual understanding of medical or curative treatment as beneficial to the patient. The deduction of a third of the period of the incarceration period to obtain the criminal's consent to go through this curative treatment, stresses the problematic nature of the prosecution's entire approach. If the criminal agrees, he benefits twice: once by curing his problem, and secondly by a shorter imprisonment; if he refuses he is also punished twice: once by his not being cured, and secondly by a longer incarceration. Thus the offered reward is unreasonably compensated. This set-up is inconsistent both with our concept of punishment as well as with our concept of medical treatment and of rehabilitation.

The prosecution can still argue that we could think of this kind of treatment (which is essentially the suppression of the offender's sexual drive) as a sort of punishment. If we accept this argument then we encounter a new version of the old kind of punishment of physical castration, which was removed long ago from most of the world's legal systems. Even if the offender consents to this, it is doubtful whether society wants to return to these kind of physical punishments through the back door. And if we consider the curative treatment as being in the patient's best interests and rehabilitation, then it should not be considered as a punishment of any sort. Thus, there is no reason to link the curative treatment with the length of the imprisonment. If there are ulterior motives (such a process).

To sum up this example, I want to say that without adopting a stance towards the rehabilitative approach to punishment, I feel that such programs should be separated from the judicial process and definitely from the prosecution as a legal authority. The prosecution should sue and convict the offenders, and the social and welfare authorities should attempt to rehabilitate them.

Chapter 12

PUNISHMENT AND DOMESTIC VIOLENCE

Domestic violence is not just a sub-category of violence in general. It is inherently complex because there is an unclear border or demarcation between tough "education" within the family (disciplining children) and violence (beating children); the demarcation between the two is much clearer outside the home, as in schools and other institutions. Another problem is that when the offender is a parent, for example, then punishing him or her is likely to harm the family at large, such as by taking away their means of support. Generally, punishment involves singling out a specific offense and the perpetrator of that offense, and this is more problematic in a family situation. These complex issues are often used by the authorities as an excuse to abstain from pursuing and punishing offenders. Consequently, they renege on their two obligations—to punish criminals and protect the weak.

In this chapter, I deal with the problem of mild punishments for violent crimes, which is reminiscent of the similar issue of the previous chapter (mild punishments for sexual crimes). The extent of intrafamilial or domestic violence, and the horrific levels of cruelty involved, probably point to a worldwide epidemic.

The second problem, more specific to Israel, is the intolerable laxity with which this phenomenon is treated by institutions and governmental institutions in Israel, including the welfare systems but most notably—the legal and law enforcement systems. In general, the legal system in Israel exhibits outrageous clemency towards criminals (such as relatively light sentences for murder), and mild punishments for domestic violence.

Domestic violence, particularly when it is aimed at children, is complicated for the reasons cited above: there is a thin line between discipline/punishment and actual violence. Hamner and Turner argue that "in our society the terms *discipline* and *punishment* are used synonymously. When someone says, 'what the child needs is discipline,' usually it means that the child needs punishment, most often physical in nature."[1] The authorities often hesitate to label punishment within the family as domestic violence or a crime or felony, particularly among people who still consider physical punishment as a legitimate means to achieve discipline within the family. For example, if a parent spanks their own child then the authorities will probably not respond, whereas they are likely to do so if the same people spank someone else's child. This is because a family has a degree of comprehensive autonomy regarding intrafamilial relations and the education (or "imposing discipline") of the children within the family.

In any event, I mainly deal in this chapter with the harsh cases of domestic violence—those labeled as crimes, some of which are even prosecuted and judged in criminal courts—and not borderline cases. These criminal cases were considered by Straus, Gelles, and Steinmetz as belonging to the category of "abusive" violence. Acts of this kind "carry a high risk of serious injury and which, if carried out between persons who are not members of the same family would be considered a criminal assault."[2] Such assaults between husbands and wives occur in the US at an annual rate of 6% of American couples (and with the allowance of underreporting, we might even double that figure).[3] The figures are less definitive in Israel, though women's organizations estimate the number of Israeli battered women to be between 200,000 and 300,000 out of an entire population of about 7,000,000 (This figure, unlike the American figure, includes battered women who do not turn to the police). The situation in the UK is also grim. Jon Bernardes quotes Anthony Giddens, saying: "The home is, in fact, the most dangerous place in modern society. In statistical terms, a person of any age or of either sex is far more likely to be subject to physical attack in the home than on the street at night. One in four murders in the UK is committed by one family member against another."[4] These grim statistics show that domestic violence has become a worldwide epidemic.

Steinmetz summarized some studies about *marital violence* (mainly in the US), and her data are shocking. She says that "between 50% and 60% of the couples interviewed, reported physical violence by a partner at some time during the marriage or relationship."[5] She also cites that "For about 1 out of 5 women,

---

[1] Hamner Tommie J. and Turner Pauline H. *Parenting in Contemporary Society*. Allyn & Bacon, Boston, 1996, p. 46.

[2] Straus Murray A., Gelles Richard J., and Steinmetz Suzanne K. "Physical Violence in a Nationally Representative Sample of American Families." In: Trost Jan. (ed.), *The Family in Change*, International Library, Vasteras, 1980, p. 163.

[3] These data are taken from Straus Murray A., ibid.

[4] Bernardes Jon. *Family Studies*. Routledge, London, 1997, p. 72. This quotation is from Giddens A. *Sociology*. Cambridge: Polity, 1989, p. 408.

[5] Steinmetz Suzanne K. "Family Violence: Past, Present and Future." In: Sussman Marvin B. and Steinmetz Suzanne K. (eds.), *Handbook of Marriage and the Family*, Plenum Press, New York, p. 732.

Punishment and domestic violence

the abuse is not an isolated incident but occurs repeatedly, and for 1 out of 15 to 20 women, severe physical battering occurs."[6] With regard to *parent–child violence* the situation is more horrific. She cites studies that indicate "that between 84% and 97% of all parents have used physical punishment at some point in their child's life. A high percentage of these families continue to use corporal punishment for disciplining their children until the tenth grade . . . and the twelfth grade."[7] Steinmetz defines violence as "an act carried out with the intention of, or an act perceived of having the intention of, physically hurting another person. This 'physical hurt' can range from a slap to murder."[8] What makes the discussion of domestic violence very complicated is the fact that "acts of physical violence between members of the same family are an age-old phenomenon. No one knows if such violence is more frequent now than in the recent past or less frequent,"[9] as Straus, Gelles, and Steinmetz claim. Thus we cannot assess whether the current awareness of this phenomenon, and efforts to eradicate it, have had beneficial or detrimental results since we cannot really estimate the rates of intrafamiliar violence in the past.

Regarding the issue of battered women, there are even theories explaining how women's compliance to being beaten provides the feedback that reinforces their husbands' violent behavior. Cheal brings Giles-Sims study below:

> "At the most general level, women's commitment to marriage causes them to sustain their husbands' violence through their maintenance of the family system, within which violent behavior is established. More specifically, wives are also thought to reinforce violent behavior when their compliance allows the abusive husband to achieve his goal. Wife beating, that is to say, is rewarded by the wife's compliant response, or in other words she provides 'positive feedback' for his violent behavior . . . Externally the major factor in Giles-Sims . . . account of wife beating is the lack of significant 'negative feedback' from outsiders . . . Battered wives . . . are described as actively shaping the impermeable boundaries of their family system. The stigma associated with family problems in general and family violence in particular leads such women to hide their situation from others."[10]

Such controversial theories, if true, only strengthen the demand for external "negative feedback" from outsiders, assuming that such feedback is necessary to help battered women to escape their seemingly endless circle of misery and violence. And a significant feedback of that kind, as I will show later in this chapter, can be the Courts' decisions.

---

[6] ibid.
[7] ibid, p. 736.
[8] ibid, p. 729.
[9] Straus Murray A., ibid, p. 149.
[10] This quotation is from Cheal David. *Family and the Stare of Theory*. University of Toronto Press. Toronto, Toronto, 1991. p. 78.

In any event, Straus, Gelles, and Steinmetz claim that "the family is truly the most violent civilian institution in American society. This situation probably characterizes families in many other societies,"[11] and this claim is likely true for the Israeli society as well. They also cite a number of factors that probably contribute to intrafamiliar violence in the US, factors that are probably relevant to the Israeli society as well. These are: unintended training in violence stemming from reliance on physical punishment; the use of force to maintain male dominance in the family; the high level of violence in the society; the conflicts which are inherent in intimate groups such as the family; and the fact that, as children, millions of husbands and wives observed violence by their own parents toward each other, something which serves as a powerful role model for their own behavior as adults.[12]

Violence in Israeli society today differs from violence in other Western countries, by the fact that the rate of violence in Israel had been relatively low for a long time and has sharply escalated over recent years, due to many local factors. Israeli society suffers from the stress of living in the Middle East with security, terrorism, and economic problems. A relatively high percentage of Israeli citizens have been subjected to the violence of terrorism over the last 4 years, and some psychologists claim that as a result, a high percentage of its citizens suffer from symptoms of post-traumatic stress disorder. This, in turn, is likely to raise the rates of violence within the society. In addition, Israel contains a relatively large percentage of recent immigrants (mainly from the Former Soviet Union and Ethiopia)—approximately 15%—and this population is typified by many of the problems of first-generation immigrants in a new country. All of these factors are likely to contribute to escalating rates of domestic violence as well as the high levels of brutality involved in such violence.

Why, then, do I raise the issue of this phenomenon in Israel, when it is so ubiquitous in other Western societies? The reason is because the standards of punishment in Israeli courts for crimes of domestic violence are extremely lower than in the United States. This does appear to obviate the utilitarian view of punishment, since it does seem that harsh punishments have not been effective in deterring intrafamilial violence in American society. However, my point is that the mild punishment for such crimes is a dangerous phenomenon in and of itself, beyond the utilitarian worldview.

The phenomenon of mild punishment for domestic violence exhibits a kind of collaboration of the legal system with the general public attitude of condoning intrafamilial violence. It is also should be seen within the context of the overall problem of the widespread leniency of Israeli court punishments for all types of violent crime, including murder—a practice that has permeated the legal system and courts for many years. Sentences that are already mild are often mitigated

---

[11] Straus Murray A. ibid, p. 163.
[12] ibid.

even further in verdicts involving domestic violence. This should enrage and infuriate all decent, law-abiding citizens.

In the previous chapter, I discussed the phenomenon of the relatively mild punishments which are imposed on sex criminals. Since many (if not most) cases of domestic violence also involve sexual attacks or incest, it may be that the mild punishments imposed for such domestic crimes, are related to the mild punishments imposed for sexual crimes. However, an explanation that relates more specifically to domestic violence is one that relates to the unique status and social role which is ascribed to the family as a social institution. In the paragraphs below, I elaborate this argument.

The scandalous clemency towards intrafamilial violence is the result, first and foremost, of the special nature of intrafamilial relations as perceived by society. Such relations are based on a unique kind of obligation that is not only the result of contractual relations, but of relations of love, affection and often even self-sacrifice or altruism (parents deprive themselves to provide for their children). These relations endow a unique, socially acknowledged status to the intimate framework of the family. A significant component of this status is the wide range of autonomy accorded to the family by society, an autonomy that is usually respected by the State through minimal interference in what goes on within the family. Part of parental autonomy is their privilege to make decisions affecting the children's future, a privilege which result from their being responsible for the children's welfare and development. On the other hand, society still maintains that parents may be sued in court for criminal offense in neglecting certain basic obligations and duties. For example, parents who do not feed their children (causing malnutrition), or those who leave young children unsupervised can be accused of desertion which is a criminal offence in most countries.

Part of the autonomy of the family is sometimes understood by the State and social authorities as including a form of immunity against public scrutiny and regulation. Those who upheld this view, consider the family as a private institution and thus "ought not to be subject to State supervision and control."[13] Archard reveals the fraud behind this presumption by insisting that the private sphere is subjected to the same legal prohibitions as the public sphere, and thus, "When a husband physically assaults his wife in the 'privacy' of their shared home, he is as guilty of committing a crime, as if he had attacked her in a 'public' space. The sexual abuse of a child by her father is no less of a crime for being perpetrated in the family home."[14] Those who ascribe ultimate privacy to the family, misinterpret the idea of privacy within family, and enable the abuse of this value of family life.

The misleading myth about the family as a private institution has a long history, and is probably a major reason for the difficulties in uprooting the phenomenon of

---

[13] Archard David. *Children, Family and the State*. Ashgate. Aldershot. 2003. p. 123.
[14] ibid, p. 124.

domestic violence. Hutchinson argues that "For cultural and political reasons the family has been regarded as a sacred institution, representing domestic tranquility. The *Journal of Marriage and the Family* contained no indexed references to violence from its first publication in 1939 through 1969... Child abuse, murder, incest, rape, and spousal abuse were non-events. Violent acts between family members were considered a private affair, if not legitimate norm."[15] A sorrowful consequence of this secrecy created the situation where 45% of women who had experienced domestic violence in London did not even tell anyone, and only 14% of experiences of abuse were reported to the police.[16] And this secrecy opens the gate to abominable crimes, whose perpetrators make use of society's conspiracy of silence in order to escape from being caught and prosecuted. The aegis of public silence and negligence made domestic violence a crime with minimal risk, and thus, almost worthwhile.

The unique status and authority endowed to parents within the framework of the family, often serve to prevent interference of social and legal authorities in what goes on within the family. Parents are generally allowed to raise children according to their best judgment. The source of this tradition is probably the Roman society which "allowed the head of the family *(Paterfamilias)* wide powers of regulation and control over his family unit, with minimal interference by the legal system."[17] The current norm which gives husbands and fathers such authoritative status, might legitimize domestic violence in the eyes of many husbands and fathers as a modern version of Roman approach.

In many cases where parents act violently towards their children or their protégés, the authorities tend to justify such acts on the grounds that parents have the authority to educate and discipline their children as they see fit. Another excuse is the parents' status as the breadwinners or the maintainers of the children—something which entitles them to a sort of "ownership" or "possession" of the whole family. This "ownership" plays a significant role in the lenient or condoning attitude (both by society as well as by the legal system) toward intrafamilial violence, on the grounds that owners have the right to treat their possessions any way they want. Such a viewpoint could influence a court to accept the assertion of a man who murdered his wife because she was "disloyal" or that she "betrayed" him. This often succeeds in changing the status of the murder from first-degree murder to killing/manslaughter, because the woman's behavior "provoked" the man.

There are many theories which use the terms of "property" or "ownership" in order to describe family life and relations. Bernard Farber, for example, speaks

---

[15] Hutchinson Janice. "Battered Women." In: Spurlock Jeanne and Robinowitz Carolyn B. *Women's Progress: Promises and Problems.* Plenum Press, New York, 1990, p. 174.

[16] These details are taken from Bernardes Jon. Ibid, p. 74, where he quotes the relevant researches which found these findings.

[17] Harding Christopher and Ireland Richard W. *Punishment: Rhetoric Rule and Practice.* Routledge, London, 1989, pp. 48–49.

about "the concept of the family as a system of property rights,"[18] and considers that in the family, "people are the property of their kinship units and can be either valued for themselves or used in exchange."[19] Jacob Joshua Ross expresses the idea of ownership in a more explicit utterance. He says: "in conceiving a child, parents have established a primary right of parental ownership in the fetus, which expresses itself, upon the child's birth, as a primary right of parental ownership in the child. The child belongs to them, and conversely they belong to the child."[20] However, when the idea of a family as property is misunderstood and abused by fathers or husbands, they may treat their children or spouses as commodities or property and not as human beings.

This causes David Archard to argue that the family's immunity against external control is conditional. He says:

> "The normal everyday activities of a family may be presumed to be such that they do not warrant the attention of public officials. This presumption is defensible. If it can be shown that what is going on in the family merits official interest, no appeal can be made to the essential privacy of family life to block that interest. What is going on merits official interest if the behavior of family member crosses a specified threshold beyond which it is deemed unacceptable."[21]

Archard claims that the unique status of the family, which entitles it to autonomy and privacy, is valid only when the family functions according to the social expectations of this institution. These expectations hold that the aims of family life would be better achieved if the family was ascribed privacy and autonomy, and that this autonomy is appropriate for the raising of children. However, this autonomy is always subject to the appropriate use and fulfillment of these values. When this autonomy is abused, then the family is no longer entitled to autonomy and privacy.

Archard explains the mandate to avoid misuse or abuse of the unique status of the family below:

> "For instance, and most centrally, the liberal state leave parents free to raise their children as they see fit. It does not do so because it views parenting as an essentially private activity with which it has not legitimate concern. Rather, it presumes that a parent's normal quotidian actions in respect of her child are not such as to merit official interest. It trusts that parents will devote themselves to their child's interests and safeguard its welfare. However, if it can be shown that what a parent is doing to her child crosses the threshold, then the state is not debarred from interfering in the family's life by the consideration that it is thereby trespassing upon a private domain."[22]

---

[18] Farber Bernard. *Family & Kinship in Modern Soceity.* Scott, Foresman and Company, Glenview, 1973, p. 23.
[19] ibid, p. 10.
[20] Ross Jacob Joshua. *The Virtues of the Family*. The Free Press, New York, 1994, p. 157.
[21] Archard David. ibid, p. 124.
[22] ibid.

Archard's argument is that the family's authority with regard to its members is conditional to its proper functioning as a loving, supportive, and caring unit. When these conditions do not exist, the privileges which are bestowed on the family are canceled and intrafamilial relations should then be treated as other non-familial relations within the State that are subject to State control and interference. In other words, the grounds for the unique status of the family are revoked as soon as intrafamilial violence rears its ugly head within the family. Thus, the unique status ascribed by society to the family, including the parents' autonomy and immunity against external invasion, is no longer valid. Once a family's function and role is changed from being a source of security and confidence to a source of violence and suffering, then that family is no longer immune to control by the State.

In fact, I believe that intrafamilial violence should be judged with even more severity than other cases of violence that never do not posses any immunity or special status in the first place. This demand to punish a violent father or husband more severely than a violent stranger is analogs to the demand to punish (for example) a policeman who raped a woman under his arrest, much more severely than an ordinary rapist. This is because the policeman abused and betrayed the special status and authority accorded to him to enforce law and order, making his crime especially heinous. Similarly, when parents misuse their special privileges and authority by harming their children, thus betraying their parental role, they should be punished even more severely than strangers who committed violent crimes against unrelated children.

Another reason to be strict about intrafamilial violence is essentially based on considerations of values and norms, as expressed in the previous chapter. Society expresses its attitude towards a particular crime by the severity of the punishment it dispenses for that same crime: a severe penalty is dispensed for a severe crime, and a mild penalty is meted out for a less severe infraction. Thus, mild punishments for domestic violence serve to reflect society's condoning attitude toward domestic violence. However, as I said in the previous chapter, sometimes punishment should not only reflect social norms but also determine or establish those norms and values. More stringent punishment should be imposed on intrafamilial violence in order to express the clear message that those who abuse their status as parents, undermine the whole institution of the family that is crucial for the existence of society as whole. Such parents (or other family members) undermine the trust bestowed upon the family by society, and abuse their authority over their families that was ascribed to them by the social framework, in much the same way as the rapist policeman abused his authority. Thus society must respond severely against this phenomenon, even as an act of self-defense.

This demand for social intervention as an act of self-defense can be derived from David Morgan's idea that "family structures might help to ensure the success of the society as a whole."[23] Winfield goes even further and argues "the free family is a precondition of civil society and political democracy independently of

---

[23] This is David Cheal's interpretation to David Morgan's theory. See, Cheal David. ibid, p. 36.

its role in upbringing."[24] If the institution of the family is so significant to social functioning then its preservation is a primary interest of society, and thus it has to intervene in order to preserve societal interests.

However, there is another significant reason for society to intervene in family life when this framework does not function appropriately. The immunity bestowed by society on intrafamilial relations makes society itself accountable and even responsible for the safety and security of family members. In the case of children, the State is also their *Parens Patriae* and as such is responsible for their security and welfare when these are not achieved within their families. In fact, abstaining from our duty to punish domestic violence more stringently, might impose the responsibility for the consequences and for the suffering of the victims on Israeli society as a whole. This might bring Israel closer to Homer's description of the Cyclopes, where "There is neither law nor public meeting, but they all live on the top of high mountains in hollow rocks, and each judges his children and woman as he pleases and pays no heed to the others."[25] What made this situation possible was that there was no legal or civil system to control these giants.

To sum up this issue, I want to reiterate several crucial considerations in dealing with domestic violence, and possible solutions to reduce its scope. The first issue, expressed by David Archard, is the social acknowledgment that domestic violence should be regarded as every other kind of assault:

> "Of course, a crime is a crime wherever it is committed. Assaults upon the person are criminal even if the other person is a family member. Nevertheless, I grant that what constitutes an assault may be thought contentious. Smacking a stranger in the street is a criminal assault. Smacking one's child in the home may be thought merely to be reasonable paternal chastisement. However, expressing things this way reveals the oddity of believing that there is a relevant difference between the two actions. Not counting as an assault on a child what would count as an assault if done to an adult surely displays our inconsistency in prosecuting crimes against the person. It does not show that there is a relevant difference between an unacceptable assault and acceptable punishment."[26]

The demand that violence between family members should be treated as violence between strangers, relates to all intrafamilial relations including those between husband and wife. Beating one's wife should be regarded as a physical assault which should be treated as it is, namely as a legal crime. And, in my opinion, this crime should be punished even more severely than the case of ordinary violence, as I explained above.

---

[24] Winfield Richard Dien. *The Just Family*. The State University of New York Press, Albany, 1998, p. 197.
[25] This quotation is taken from Henting Hans von. *Punishment: Its Origin, Purpose and Psychology*. Paterson Smith Publishing Cooperation, 1973, p. 29.
[26] Archard David. ibid, p. 125.

The second issue I want to stress is the need for State intervention even in borderline cases or those belonging to the "twilight zone," such as in dysfunctional families where the children are neglected, even if it does not constitute criminal neglect. Even though actual crimes have not yet been committed, state intervention would help to prevent a legal crime (such as criminal neglect), but also to intervene and help the family to improve its functioning. I fully agree with Archard's view, who says:

> "Intervention into a family need not only mean the prosecution of a crime. The involvement of public agencies in a family's life may be initiated by non-criminal activity that is nevertheless thought sufficiently serious to defeat the presumption of familial privacy. Most Western jurisdictions work with something like the rule that where there is a real risk of significant harm befalling the child as a result of parental action or inaction, then (but only then) official intervention is justified."[27]

This view expresses the idea that domestic violence shares certain similarities with the progression of malignant disease: it has symptoms that should be treated as early as possible to prevent deterioration of the condition; and it tends to escalate and worsen with time if left untreated. Thus the authorities should be attentive to early warning signals of familial abuse in order to either rehabilitate the family unit, or at least preempt its further deterioration. Though this invades the family's privacy, it can maintain the welfare of its members as individuals and also, perhaps, preserve the existence of the family as an important social unit.

The last issue I want to raise here is the importance of family intervention not only by legal authorities but also institutions of the civil society, such as social workers and welfare officers. Together, these bodies should establish a supportive apparatus to enable family members to maintain stable, mutual family relations, thus reducing the circumstances which sometimes cause domestic violence. Although these bodies and agencies exist today in most Western states, they are usually underfunded; their workers are often overworked and underpaid, and thus cannot provide the supportive apparatus that is necessary.

Archard presents this as follows:

> "It would be a mistaken oversimplification to view 'official intervention' as only taking the form of the prosecution of criminal activity and the imposition of legal sanctions. A whole range of public agencies can be involved with a family—social workers, community workers, health workers, and welfare officers. Indeed, some writers insist that the modern family is more thoroughly 'policed' by such quasi-official professionals than might be suggested if one's model of governance was the operation by government of legal-coercive measures."[28]

---

[27] ibid.
[28] ibid.

The hope is that the rates of domestic violence could be sharply curtailed by a governmental framework of professionals dealing with families in distress (such as social workers and family therapists) to prevent the disintegration of such families or at least avoid their deterioration into violence. In addition, more severe punishments should be imposed on domestic violence to reflect the fact that there would be fewer excuses for this phenomenon. In other words: those who continue to treat their family members violently, despite wide-spread social help and support, would rightly be considered as criminals who would be punished accordingly. The social and normative message of such punishment would be that society makes all necessary efforts to sustain the institution of the family and its functioning, and punishes those who harm the institution or its members.

Hutchinson researched the issue of battered women and came to the conclusion that community support is crucially significant in the prevention of this phenomenon. Communities can provide this support through victim advocate projects, establishment of shelters, etc. However Hutchinson also ascribes a critical role to the legal system. She stresses that "legal and law enforcement agencies could act to change the face of abuse. Laws must be created to require police to allow for the arrest of a violent spouse. Protection from further abuse, must be provided to the woman who wishes to press charges against the abuser. The criminal justice system should be improved and expanded to make offenders responsible for their actions through prosecution."[29] I can only share her conclusions and hope that the Israeli legal authorities will adopt these recommendations. Another idea which I warmly adopt from Hutchinson's study, is that the only way to rehabilitate society from the malignant disease of domestic violence is comprehensive cooperation between all the social and governmental institutions that deal with efforts to eradicate this social curse. No one agency alone can vanquish this epidemic and they must work together for maximum effectiveness. I choose to end this chapter by quoting Hutchinson's persuasive demand for a sincere struggle against domestic violence in general, and spousal assault in particular. Hutchinson says:

> "The primary institutions of society, family, medicine, religion, and criminal justice must send a message that violence toward spouses will not be tolerable. Representatives of all service agencies must be sensitive to the impact that their own male-dominated systems have on women. The concept of a family must be redefined to connote "partnership instead of ownership". Only when women and men can respect themselves and each other, will family members and others realize their complete and total potential."[30]

So be it!

---

[29] Hutchinson Janice. ibid, p. 186.
[30] ibid.

Chapter 13

CAPITAL PUNISHMENT AND THE MENTALLY RETARDED

In this chapter I focus not on the general debate regarding capital punishment, but on applying the death sentence to mentally retarded or mentally ill criminals.[1] The official position of the US Supreme court on this issue is as follows: "In 1986 the Supreme Court ruled that execution of the insane was unconstitutional under the Eighth Amendment, based on the nation's 'common law heritage' in which the execution of the insane was considered 'savage and inhuman'."[2] A similar attitude toward this issue was expressed during the 1980s by the *Safeguard Guaranteeing Protection of the Rights of Those Facing the Death Penalty*, which was adopted in March 1984 by the *UN Committee on Crime Prevention and Control*. "The 1984 'Safeguards' has excluded the insane from the death penalty; the 1988 resolution added 'persons suffering from mental retardation or extremely limited mental competence'."[3] William Schabas mentions that international law excluded certain categories of people from the death penalty, such as: persons under eighteen, pregnant women, the elderly, young mothers, the insane, and mentally handicapped.[4]

---

[1] I use the phrase "mentally retarded," which together with the term "mentally deficient" became familiar in North America, while in Britain the more familiar term is "mentally handicapped." On the differences between the different phrases see Byrne Peter. *Philosophical and Ethical Problems in Mental Handicap*. Palgrave Macmillan, Hampshire, 2000, p. 17.
[2] Vila Bryan and Morris Cynthia. *Capital Punishment in the United States*. Greenwood Press, Westport, 1997, p. 241.
[3] Schabas William A. *The Abolition of the Death Penalty in International Law*, Second edition. Cambridge University Press, Cambridge, 1997, p. 166.
[4] Ibid, p. 120.

There are more examples of similar practices in the United States, such as found in the Model Penal Code of The American Law Institute, section 210.6 (Sentence of Death for Murder; Further Proceedings to Determine Sentence), section 1 (Death Sentence Excluded). This code states that when a defendant is found guilty of murder, the court should exclude the death sentence if (paragraph E) the defendant's 'physical or mental condition calls for leniency.' In section 4 (Mitigating Circumstances), paragraph B, 'mitigating circumstances' are to be considered for a murder committed while the defendant was under the influence of extreme mental or emotional disturbance.[5] Different states, such as Arkansas, Colorado, Florida, and Nebraska "specified mitigating circumstances that sentencers were required to weight against aggravating factors, such as mental impairment, youth, and extreme emotional disturbance at the time the crime was committed."[6] Other states that have statutes forbidding execution of the mentally retarded are Georgia, Indiana, Kansas, Kentucky Maryland, New Mexico, New York, Tennessee, Washington, and the federal system.[7]

However, although it seems fairly obvious that mentally ill and mentally retarded offenders should not receive the death penalty in American courts, in point of fact they sometimes do receive verdicts of capital punishment, particularly in Texas. In this chapter I explain why I think that the imposing of such a punishment on mentally retarded or insane criminals is morally wrong.

R. J. Wallace connects the notion of responsibility to that of accountability. He says:

> When we ask whether a given person is morally accountable for a given action, we are asking whether it would be fair to hold the person responsible for the action. Holding responsible is in turn an attitudinal stance that we adopt toward a person, involving a susceptibility to reactive emotions on account of what the person has done, and a disposition to respond to the person's in ways that express the reaction emotions. Thus, a person is blameworthy for having done something morally wrong when it would be fair to feel resentment or indignation in response of the person's conduct, and to engage in sanctioning behaviour that express such reactive sentiments.[8]

Wallace explains that the basis for accountability is "deliberative authority," which in turn has two aspects. The first aspect is whether a given agent has enough reasoning and insight to understand and comply with moral demands (such as not to steal or murder). The second aspect is whether an agent possesses the required measure of reason to comply with moral demands; whether the agent

---

[5] These details were taken from: Zimring Franklin E. and Hawkins Gordon. *Capital Punishment and the American Agenda.* Cambridge University Press, New York, 1986, pp. 79–80.
[6] Haines Herbert H. *Against Capital Punishment.* Oxford University Press, 1996, p. 46.
[7] These details were taken from: O'shea Kathleen. *Women and the Death Penalty in the United States, 1900–1998.* Praeger, Westport, 1999, p. 31.
[8] Wallace R. J. "Reason and Responsibility." In: Cullity Garrett and Gaut Berys. (eds.), Clarendon Press, Oxford, 1997, p. 321.

has *most* of the reasoning skills necessary to comply with moral demands, and whether the agent's compliance with moral demands is *optimal* from the standpoint of deliberative reason.[9] The conjunction of these two measures determines the agent's general competence or capacity for reflective self-control, and this competence can vary on a scale from developed to impaired. When competence is substantially impaired, we should adjust our judgment of moral responsibility accordingly. Wallace argues that "cases in which the powers of reflective self-control are substantially impaired, or in which it is extremely difficult for an agent to exercise those powers, should be treated as cases in which the agent's moral accountability is diminished."[10] The concept of diminished moral accountability will be elucidated in this chapter and translated to the concept of diminished legal responsibility and accountability. I argue that this notion of diminished responsibility should preclude the imposition of capital punishment on mentally retarded and mentally ill people, without entering into the ideological dispute regarding capital punishment per se. What I want to stress is that even if we accept capital punishment as an appropriate response to homicide, it should not be applied in cases of mentally retarded and mentally ill offenders.

The issue of punishing mentally retarded offenders is itself very controversial. R. A. Duff, for example, thinks that offenders who lack the capacity to realize that imprisonment is a punishment, are not fit to be punished. "For punishment aims, and must aim, if it is to be properly justified, to address the offender as a rational and responsible agent: if she cannot understand what is being done to her, or why it is being done, or how it is related as a punishment to her past offence, her punishment becomes a travesty."[11] The tendency not to indict such offenders is supported by the Report on the Committee on Mentally Abnormal Offenders (the Butler Report), which "recommended that a special verdict of 'not guilty on evidence of mental disorder' should be returned 'if at the time of the act or omission charges the defendant was suffering from severe mental illness or severe subnormality."[12]

For the purposes of our discussion here, it makes little difference whether we take the stance of psychiatrists who consider mental illness to be a disease, or that of sociologists who regard mental illness as a behavior that violates social norms. Following Bernard J. Gallagher III's claim that "It is exceedingly difficult to define mental illness objectively,"[13] I choose to sidestep the issue. I content

---

[9] Ibid, p. 322.
[10] Ibid, p. 323.
[11] Duff R. A. *Trials & Punishment*. Cambridge University Press, Cambridge, 1986, p. 27.
[12] This quotation is from Ten C. I. *Crime Guilt and Punishment*. Clarendon Press, Oxford, 1987, p. 126. There Ten quotes from: Home Office, Department of Health and Social Security, *Report of the Committee on Mental Abnormal Offenders*, HMSO Cmnd. 6244 (London, 1975), para. 18.30.
[13] Gallagher Bernard J. III. *The Sociology of Mental Illness*. Prentice Hall Inc, Englewood Cliffs. 1980, p. 33.

myself with the assumption, supported by Herbert Fingarette, that "mental disorders are indeed objectively existing conditions, not 'myths' as claimed by the so-called antipsychiatry movement."[14] This means that I consider mental illness as a state that can be identified professionally, and that there is comprehensive agreement regarding the harm it has on person's cognitive judgments. I accept Allan V. Horwitz' model which argues that "It is most appropriate to regard characterization of mental illness as a distinctive type of reaction to undesirable behavior,"[15] and that "The most obvious cases of mental illness are characterized by the appearance of incomprehensibility."[16] For our purposes, it is not necessary to define or label the phenomenon of mental illness but only to relate to its main epistemological characteristic of incomprehensibility. This characteristic has ethical implications, particularly to the extent to which we regard mentally ill people as responsible for their actions and accountable both morally and legally.

It is often the case that a court may deem mentally impaired persons as being fit to stand trial, even when there are doubts regarding their full liability for their actions. Mental illness or retardation may affect offenders' conduct in some areas of their lives and not others, and they can be held responsible for at least some of their acts.[17] C. I. Ten, for example, says that a mentally ill offender may be guilty of a crime that was not rooted in his mental illness but sprung from criminal motives. On such occasions we cannot exempt mentally retarded offenders from being tried or even being punished. However, the person's mentally retardation or illness must endow them with some measure of diminished responsibility for their actions. This has to be taken into consideration during the trial, and definitely when deciding on the sentence. For example, we might impose a diminished punishment on these offenders. In my opinion we should not impose capital punishment, the most severe penalty for homicide in our scale of punishment, on a mentally retarded or mentally ill offender. Imposing capital punishment revokes the entire concept of diminished responsibility and undermines the connection between rationality and responsibility, and between responsibility and accountability. By doing so, we jeopardize the very rationale for punishment as expressed above by R. A. Duff. The example below illustrates how the whole idea of diminished responsibility is sometimes ignored by US Courts.

CASE 1

This example deals with the case of John Paul Penry (born April 5, 1956), who had an estimated IQ of between 50 and 63 and the mind of a 7-year-old child. He

---

[14] Fingarette Herbert. *Mapping Responsibility*. Open Court, Chicago and La Salle, 2004, p. 55.
[15] Horwitz Allan V. *The Social Control of Mental Illness*. Academic Press, New York, 1982, p. 25.
[16] Ibid.
[17] See Ten C. I. Ibid, p. 126.

suffered from cerebral damage at birth and was abandoned by his mother (and nobody has ever heard of his father). He grew up in homes for abandoned children, where his mental retardation was worsened. In 1979, he raped and murdered a 22-year-old woman, Pamela Moseley Carpenter in her home. The victim managed to describe the attacker before she died, and Penry, who fitted that description and had a previous conviction for rape, was arrested and confessed that he committed the crime. Penry's IQ, mental age and social maturity (of a 9 or 10-year-old), together with "other mitigating evidence pertaining to Penry's mental retardation, learning disabilities, and childhood abuse, had presented during Penry's trial but had not influenced the jury's decision to impose the death penalty."[18] This decision was considered unreasonable and the case rose to the US Supreme Court. "Penry's death sentence has been overturned by the US Supreme Court twice, and twice Texas has re-sentenced him to death. The second time the Supreme Court ruled it unconstitutional to execute the mentally retarded; a Texas jury then found that Penry was not retarded and thus eligible after all for the death penalty. He is still on death row."[19]

Penry's defense claimed that his retardation rendered it unconstitutional to execute him according to the Eighth Amendment to the US constitution, which requires that "Excessive bail shall not be required, nor excessive fines imposed, nor cruel and unusual punishments inflicted." However, the Court's opinion, delivered by Justice O'Connor, was that since Penry was found competent to stand trial by the jury, his claim was doubtful; or in O'Connor's words, whether "any mentally retarded person of Penry's ability convicted of a capital offense [can be exempted from the death sentence] simply by virtue of his or her mental retardation alone."[20] In any event, the Supreme Court overturned Penry's death sentence. Justice Brennan who wrote the partly dissenting opinion (together with Justice Marshal) said that "Because I believe that the Eighth Amendment to the United States Constitution stands in a way of a State killing a mentally retarded person for a crime for which, as a result of his or her disability, he or she is not fully culpable, I would reverse the judgment of the Court Appeals in its entirety."[21]

On the surface, it appears that the dispute between Penry's advocates and the Texas Court, and even between the US Supreme Court and the Texas Courts, hinges on the question regarding whether Mr Penry is really mentally retarded or not. However, this is not the real story with regard to the insistence of the Texas Court on the execution. Other analogous cases of execution of mentally retarded or insane criminals as carried out by Texas courts, reveal the attitude, preference,

---

[18] Vila Bryan and Morris Cynthia. Ibid, p. 241.
[19] http://bipolar.about.com/od/socialissues/a/040628_texas.htm.
[20] Justice O'Connor which delivered the opinion of the court (492 US 302 1989). Quoted from Vila Bryan and Morris Cynthia. Ibid, p. 243.
[21] Justice Brennan who wrote the dissenting opinion (492 US 302 1989). Quoted from Vila Bryan and Morris Cynthia. Ibid, p. 244.

and tendencies of these Courts is general, and toward mentally retarded offenders in particular. I cite a few illustrative examples below.

Oliver David Cruz, 33, was executed for the 1988 abduction, rape and fatal stabbing of a 24-year-old woman stationed at Kelly Air Force Base in San Antonio. Cruz's IQ tested as low as 63, leading death penalty opponents to argue that he should not be executed. The Supreme Court, which has allowed other mentally ill or retarded inmates to be executed before that case, voted 6-3 to allow the execution to go on. Oliver David Cruz was executed in Texas August 9, 2000. However, Cruz was not a unique case, and there were many other cases of mentally retarded or insane people who were executed in Texas within a very short period. Here are few more examples.

Larry Robison, who suffered from paranoid schizophrenia, was executed on January 21, 2000. Gary Graham, who had been an abused and neglected child with a history of mental illness, was executed on June 22, 2000. John Satterwhite, diagnosed as a paranoid schizophrenic with an IQ of 74, was executed in August 16, 2000. Kelsey Patterson's case is another example of the propensity of Texas courts to execute the mentally ill: Patterson's paranoid schizophrenia was so severe that he was twice found not guilty by reason of insanity in previous nonfatal shootings. He was treated for his mental illness in these previous shootings and released. Then, when he was sentenced to death for a double homicide, Patterson's condition was such that the Texas Board of Pardons and Paroles even recommended to Governor Rick Perry that his sentence be commuted, but Perry refused. Patterson was executed in May 18, 2004.[22]

These examples show that the Texas courts are not much swayed by the impaired mental condition of the offenders when deciding on imposing capital punishment on them. George. W. Bush had signed many execution orders during his term as Governor of Texas and when he was first elected to the US presidency (close to the time of Cruz's execution and Penry's appeal to the Texas Supreme Court), there was an incisive public debate regarding Texas' wholesale execution policy. However, this debate did not focus on the execution of criminals with impaired cognitive abilities (4 years later, Bush was re-elected for another presidential term.)

The large number of executions in Texas might well have led Tom Campbell and Chris Heginbotham to the idea that: "In fact, we often deal more harshly [with those of impaired mental conditions] than is appropriate in the case of the fully responsible perpetrator of harms who can be said to deserve punishment."[23] This utterance reflects the results of a special investigation of the UN Commission on Human Rights, which was published on February 1996. This investigation "documenting the fact that many death sentences in the United States 'continues

---

[22] http://bipolar.about.com/od/socialissues/a/040628_texas.htm
[23] Campbell Ton and Heginbotham Chris. *Mental Illness: Prejudice, Discrimination and the Law.* Dartmouth, Aldershot, 1991, p. 101.

to be handed down after trials which fall short of international guarantees for a fair trial'. The probe cited six instances where death was sentence despite doubt about guilt. Twelve others were sentenced to death despite serious mental retardation, a violation of human rights the commission termed 'particularly disturbing'."[24] The Texas courts might be the worse examples of this phenomenon, but definitely not the only ones.

Many people who demand to exchange Penry's punishment for the life sentence or other penalty, do so because they object to capital punishment per se. They maintain that no crime can justify a criminal's execution in a civilized society and they equate capital punishment with a form of institutionalized revenge or vendetta exercised by the state under the guise of legal authorization. Thus they demand the eradication of the death sentence from the spectrum of punishments throughout the entire world, and do not discriminate among the crimes for which this punishment is imposed. Since their goal is to abolish this penalty for everyone, they do not even address the mental state of the offender. But in the Penry, Cruz, Satterwhite, and Patterson cases, even those who support capital punishment for terrible crimes should have doubts and hesitations regarding the appropriateness of imposing capital punishment on such mentally impaired criminals.

The origins of such doubts are grounded in our general attitude toward punishment. We believe, as was discussed in detail in Chapter 11, that the severity of the punishment should be proportional to the gravity of the offence. However, we also believe that significant weight should be ascribed the degree of responsibility and accountability of offenders: the offenders' extent of control in abstaining from the crimes they committed under the circumstances. Sometimes the circumstances of a specific event reduce the offender's responsibility and accountability, and thus should accordingly reduce the punishment for that offence. For example, a truck driver involved in a car accident under difficult visibility conditions on a slippery and curvy road will be considered less responsible for the consequences of that accident, than if it had occurred under optimal driving conditions. Thus he should receive a milder punishment than if the accident had occurred under better driving conditions.

When we deal with mentally retarded or mentally ill criminals our basic intuition is that it is pointless to hold responsible "those who are not responsible agents."[25] Ted Honderich maintains that when we construct a penalty system we should:

---

[24] Megivern James J. *The Death Penalty: An Historical and Theological Suvey.* Paulist Press, Mahwah, 1997, p. 451.

[25] Morris Herbert. *On Guilt and Innocence*. University of California Press, Berkeley, 1976, p. 86. Even though Morris has problems with Herbert Fingarette's claim that "so far there is insanity there cannot be responsibility," he agrees, at least in principle, that responsibility can be ascribed only for responsible agents. Fingarette's claim is on Morris Herbert. *On Guilt and Innocence*. University of California Press, Berkeley, 1976. p. 76.

reflect on actions of considerable harmfulness. The grievance to which they may be regarded as giving rise, however, is a function not only of their harmfulness but also of the extent to which the agents involved are responsible for their actions. We initially fix a particular penalty for actions of a particular harm, say killing, *given that* they are actions for which the agents can be held wholly responsible. We then fix lesser penalties for actions of this kind where the agents are to a lesser extent responsible.[26]

Honderich means that a person's culpability or guiltiness is determined by two considerations: the harm entailed by the person's action, and the extent to which the person is responsible for that action. These two components together determine what Honderich calls "the grievance" this person caused, according to which he/she should be punished.

However, a no less important part of the offender's responsibility and accountability is the extent to which the offender understood the situation and could have foreseen the possible consequences of his actions. It is clear that ethically, mentally retarded offenders bear diminished responsibility and accountability since they do not have full reasoning and consideration to understand the situation and circumstances of the crime. We definitely do not ascribe them full understanding of the tragic consequences of their actions.

If we agree with Honderich's claim that responsibility is a central component in determining punishment, then we can equate mentally retarded individuals with mentally ill ones as cited below in Herbert Fingarette's analysis. Fingarette's argument is as follows:

1. Insanity is (a) irrational conduct (b) from a grave defect of the person's capacity for rational conduct (c) which is at least for the time an inherent part of the person's makeup.

If we now substitute for the term "rational" and "irrational" the central meaning of those terms as developed on the earlier discussion [that is the capacity to respond relevantly (O.E)] we derive:

Insanity is failure to respond relevantly to what is essentially relevant by virtue of a grave defect in capacity to do so inherent at least fro the time in the person's mental makeup.

Since 1. and 2. make it plain that where there is insanity there is inherent incapacity to respond relevantly, it is evident that:

So far as there is insanity there cannot be responsibility.[27]

I believe that we can apply Fingarette's assessment of the insane regarding criminal responsibility, to that of the mentally retarded as both cannot respond

---

[26] Honderich Ted. *Punishment*. Penguin Books, Harmondsworth, 1969, p. 31.

[27] This quotation is from Morris Herbert. Ibid, pp. 75–76. He refers there to Fingarette's Herbert. *The Meaning of Criminal Insanity*. The University of California Press, Berkeley, 1972, p. 203.

relevantly to situations. Fingarette defines criminal insanity as cases where "The individual's mental makeup at the time of the offending act was such that, with respect to the criminality of his conduct, he substantially lacked the capacity to act rationally (to respond relevantly to relevance so far as criminality is concerned)."[28] Thus in my opinion, a mentally retarded individual such as Penry cannot bear the full responsibility for his actions that is necessary for the infliction of the maximal penalty.

Barbara Wootton also maintains that "mental abnormality must be related to guilt; for a severely subnormal offender must be less blameworthy, and ought therefore to incur a less severe punishment, than one of greater intelligence who has committed an otherwise similar crime, even though he may well be a worse risk for the future."[29] This is because the court determines their penalties according to their assessment of one's guilt or culpability, rendering the insane or mentally retarded criminal as less guilty than the normal adult. Thus Wootton asserts that, "since an insane person is not held to be blameworthy in the same way as one who is in full possession of his faculties... he must not be hung if found guilty on a capital charge."[30]

Paul Benson discusses the issue of moral responsibility in the context of preliminary functions:

> First, regarding someone as a morally responsible agent functions to include her within some moral community in which agents can reasonably expect each other to live up to certain moral standards. It functions to affirm the person's status as a sufficiently component moral agent. Corresponding to this function is the *exemption* from responsibility of those to whom ordinary moral expectations do not apply, such as very young children, developmentally disabled adults, adults with traumatic upbringings, and so on. These persons share limited or impaired capabilities to recognize, appreciate, and respond competently to moral claims. Others in the relevant communities are, therefore unable to expect that they regularly will be responsible to those claims.[31]

It is doubtful whether an offender such as Penry, with the cognitive skills of a 7-year-old and social skills of a 10-year old, can be considered responsible enough to be a full-fledged member of the Texas moral community and, thus, tried under due process. Therefore, we have doubts regarding the extent to which Perry is

---

[28] Ibid, p. 211.
[29] Wootton Barbara. "The Problem of the Mentally Abnormal Offender." In: Wasserstrom Richard. A. *Today's Moral Problem*, Second Edition. Macmillan Publishing, New York, 1979, pp. 460–461.
[30] Ibid, p. 461.
[31] Benson Paul. "Blame, Oppression, and Diminished Moral Competence." In: DesAutels Peggy and Urban Walker Margaret. (eds.), *Moral Psychology*. Rowman & Littlefield Publishers, Lanham, 2004, pp. 194–195.

able to understand or carry out moral commands or imperatives. If he is not able, then Herbert Fingarette would have said that "Plainly it makes no sense to punish someone for disobeying a command if that person does not have the capacity to respond to such a command. The incapacity may be physical or it may be mental. But it only makes sense to command those who do have suitable capacity to respond."[32] Although Penry's retardation did not exempt him from being tried, his retardation still limited his suitability for responding to commands; John Martin Fisher and Mark Ravizza argue that "When one is morally responsible for performing an action, one must have guidance control of action."[33] How can we ascribe guidance control to a person whose cognitive skills are those of a 6 or 7-year-old boy?

However, even if Penry was eligible to be tried, the Texas Court should consider his accountability before imposing the maximal punishment for his offense. Paul Benson below cites the second function of moral responsibility:

> A second main function of responsibility-attribution is to mark particular action or failure to act as appropriate basis for assessing agent's responsiveness to moral claims. Given that a person is regarded as a competent moral agent, the issue here is whether or not a particular piece of conduct can be used to judge the moral quality of the agent's response to specific circumstances. Functioning in this second way, investigation of responsibility draws attention to the content of the agent's intentions, the extent of her control, the state of her knowledge of the situation, and so on. Corresponding to this use of responsibility-attribution are the *excuses* for apparent failure to respond acceptably that are owed to those who are compelled to act as they do, who cannot be expected to know consequences of their acts, or who are victims of unfortunate accident.[34]

Even if the Texas Courts deemed Penry as competent to stand trial, it seems strange that they deemed him possessing full control while he committed the crime, and that he had full knowledge of the situation. In short, their imposition of the maximal penalty on Penry means that they did not view his retardation as mitigating. From the point of view of moral responsibility, this attitude is quite inconsistent since it assumes that Penry had full guidance and control over what he did, something which is not compatible with his mental capabilities.

One of the most relevant terms for discussing the legal accountability of the mentally retarded is that of *mens rea*. C. I. Ten explains that: "The law uses the term *mens rea* (a guilty mind) to refer to those mental elements of conduct

---

[32] Fingarette Herbert. Ibid, p. 36.
[33] Fischer John Martin and Rivizza Mark S. J. *Responsibility and Control: A Theory of Moral Responsibility*. Cambridge University Press, New York, 1998, p. 89.
[34] Benson Paul. Ibid, p. 195.

which are necessary for criminal conviction and punishment. For example, intentional killing is killing with *mens rea*, whereas *mens rea* is absent in accidental death."[35] In order to apply the idea of *mens rea* to mentally retarded or insane offenders, Ten cites the McNaghten Rules which were used since 1843 for establishing a defense of insanity. According to these rules, "it must be proven that the accused 'was laboring under such a defect of reason, from disease of the mind, as not to know the nature and quality of the act he was doing; or, if he did know it, that he did not know he was doing what was wrong'."[36] R. Jay Wallace who interprets these rules say that "The McNaghten Rules seem to express a common view about at least some of the conditions present in cases of insanity or mental illness that make it unreasonable to hold people accountable for what they do."[37] Ten examines the *mens rea* of mentally retarded persons in the context of their punishment and says: "A mentally ill person, who satisfies the McNaghten test, might not know the nature of his act, and would therefore lack the *mens rea* necessary for criminal liability. Mental illness involving cognitive impairments can deprive persons of the knowledge and intention relevant to the proof of *mens rea*. So ... it is important that such persons should not be convicted and punished just as other offenders who lack mens rea are not convicted or punished."[38] However, even if the assessment of an offender's *mens rea* shows that she had partial understanding of the crime situation, or that she understood to some extent that what she did is wrong, this woman cannot be considered as fully accountable for the crime, and thus, as we saw earlier in Honderich's analysis, should not punished by the maximal penalty. Capital punishments should only be imposed on those with full *mens rea* who committed a similar crime.

Tom Campbell and Chris Heginbotham share the belief held by Ten, that the McNaghten Rules can "apply to severe mental handicap and also to a person whose delusional perceptions and disordered thinking render him unable to have sufficient grasp of the situation to be able to know that his behaviour is contrary to law, provided that this incapacity arises from 'mental disease'."[39] If Penry's *mens rea* was assessed according to the McNaghten Rules, he should merit mitigating considerations that should reduce his culpability and guilt, and definitely exclude the death sentence. Campbell and Heginbotham support their view about the applicability of the McNaghten Rules to mentally handicapped criminals by quoting the *American Law Institute Model Penal Code*:

---

[35] Ten C. I. Ibid, p. 100.
[36] Ibid, p. 123.
[37] Wallace R. Jay. *Responsibility and the Moral Sentiments*. Harvard University Press. Cambridge. 1994. p. 170.
[38] Ten C. I. Ibid, pp. 130–131.
[39] Campbell Ton and Heginbotham Chris. *Mental Illness: Prejudice, Discrimination and the Law.* Dartmouth, Aldershot, 1991, p. 141.

1. a person is not responsible for criminal conduct if at the time of such conduct as the result of mental disease or defect he lacks substantial capacity either to appreciate the criminality of his conduct or to conform his conduct to the requirements of the law, and,
2. the terms 'mental disease or defect' do not include an abnormality manifested by repeated criminal or otherwise antisocial conduct [that is, psychopathy].[40]

These conditions should apply to Penry's mental condition and thus, at least, mitigate the penalty.

A trend has emerged in recent years in which many legal systems around the world have started to adopt and implement the idea of diminished responsibility and accountability while handing down verdicts and sentences. Of course, if an accused is deemed completely unfit to stand trial then this is not relevant, but often an accused who is deemed fit to stand trial is still ascribed lesser responsibility and accountability when committing the offence. In such cases the court often hands down a milder punishment than the maximal penalty allowed or mandated by law. Such a policy serves to upheld the principle of fitting the severity of the punishment to the gravity of the offence—*as well as* to the degree of responsibility and accountability, or *mens rea*, of the offender.

William Kneale mentions that during the 20th century "many psychologists have argued that the relevant notion for insanity should be defined still more widely; and ... the law has recently been altered to allow for a plea of diminished responsibility in murder trials."[41] He maintains that "responsibility" in the legal context is "liability to be called to answer of account for one's act."[42] The more accurate term for this might be accountability, which is explained as "being accountable under some rule to a determinate authority for a determinate sphere of action."[43]

The plea for diminished responsibility in this context "is not treated as a complete defence on any occasion but always as a partial excuse on the strength of which a jury may classify a killing as something less than a murder."[44] However, the idea of diminished responsibility is not only relevant for the conviction, but also, and maybe more significantly, for determining the penalty of the offender. Those who support the implementation of diminished responsibility on an insane or mentally retarded offender mean that "a man who has killed someone may be responsible in a sense that he can be called to account and yet have diminished responsibility because he can not be made to pay so big a penalty as another might

---

[40] Ibid.
[41] Kneale William. "The Responsibility of Criminals." In: Acton H. B. (ed.), *The Philosophy of Punishment*. Macmillan, London, 1969. p. 172.
[42] Ibid, p. 173.
[43] Kneale William. "The Responsibility of Criminals." In: Acton H. B. (ed.), *The Philosophy of Punishment*. Macmillan, London, 1969, p. 175.
[44] Ibid, p. 173.

whose responsibility was not diminished."[45] The idea of diminished responsibility is conceived in terms of reduced culpability, and this affects the penalty that should be imposed on offenders whose responsibility is considered to be diminished.

Barbara Wootton says that "under the Homicide Act, a defense of diminished responsibility opens the door to milder punishments than the sentences of death and life imprisonment which automatically follow the respective verdicts of capital and non-capital murder."[46] Wootton reminds us that the Homicide Act includes three facets: first, the question of responsibility must be decided prior to conviction and not after it; second, the responsibility issue must be rendered by the jury; and third, the charges involved are of the utmost gravity. These three facets have caused the relationship between responsibility and culpability to be explored with exceptional thoroughness in the context of homicide. However, Wootton insists that the principles of diminished responsibility "are by no means restricted to the narrow field of charges of homicide. They have a far wider applicability, and are indeed implicit also in section 60 of the Mental Health Act."[47] This means that the idea of diminished responsibility is relevant for any decision about punishing the insane or the mentally retarded offender, though it is most acute in homicide cases and definitely when considering capital punishment.

When Texas courts impose the maximal penalty authorized by law on mentally retarded or insane offenders (who are of diminished responsibility and accountability), they are in effect totally revoking the very concept of mitigating circumstances. In addition, these courts will not logically be able to impose life sentences (a milder punishment) on any future criminal who will be convicted for the same crimes, once they did not mitigate the sentence even for the mentally retarded or insane. In effect, they relinquish the very option of taking into consideration factors of diminished responsibility and accountability in other cases as well. Such a relinquishment weakens the linkage between the severity of the punishment and the level of responsibility we ascribe to the offender; hence, it weakens the moral justification for the practice of punishment as a whole. The very institution of punishment is grounded on the offenders' moral responsibility and accountability when we demand that they "pay their bill" to the social framework within which they live.

I conclude that even those who support capital punishment should demand that it be imposed not only on those who deserve the most severe penalty allowed by society, but also those with the highest level of *mens rea*. Thus when trying mentally retarded or insane criminals, their mental state should be considered as a mitigating factor in reducing the death penalty to life imprisonment.

[45] Ibid.
[46] Wootton Barbara. Ibid, p. 461.
[47] Ibid.

I would like to summarize my stance regarding the punishment of mentally retarded or insane criminals by quoting from Justice O'Connor's opinion of the Court in Penry's case:

> It well settled at common law that "idiots", together with "lunatics", were not subject to punishment for criminal acts committed under those incapacities.
>
> The common law prohibition against punishing "idiots" for their crimes suggests that it may indeed be "cruel and unusual" punishment to execute persons who are profoundly or severely retarded and wholly lacking the capacity to appreciate the wrongfulness of their actions. Because of the protections afforded by the insanity defense today, such a person is not likely to be convicted or face the prospect of punishment... Moreover, under Ford v. Wainwright... someone who is "unaware of the punishment they are about to suffer and why they are to suffer it" cannot be executed.[48]

Penry has not been executed yet, thus giving the Texas Court another chance of preventing at least one injustice.

---

[48] Justice O'Connor which delivered the opinion of the court (492 US 302 1989). Quoted from Vila Bryan and Morris Cynthia. Ibid, p. 242.

INDEX

abortion, 17, 76, 84, 91, 115, 128
academia, 2, 3, 37, 38, 47
academic freedom, 2, 38, 44, 45, 47, 48
accountability, 15, 143, 170–172, 175, 176, 178, 180, 181
active euthanasia, 60, 61
adoptions, 40, 98, 99, 112, 113, 118, 121, 122, 125, 127–129, 131–140
adult stem-cells, 82
adulthood, 7, 82, 122
African Americans, 19
Agar, N., 83
AIDS, 9
Allen, Richard and Vickie, 133
allowing someone to die, 56, 57, 60, 61, 63
Almond, Brenda, 16, 79
Alpern, Kenneth D, 128, 124, 125, 134, 136, 138, 139
altruism, 71, 74, 78, 80, 161
Alzheimer's, 82, 84
American Law Institute Model Penal Code, 170, 179
American Medical Association (AMA), 60
anemia, 84, 91
anesthesia, 78
anonymity, 8, 98, 111–114
Antinori, Severino, 89
Archard, David, 101, 102, 107, 108, 110, 117, 127, 129, 161, 163–166

Arendt, Hannah, 41
Arkansas, 132, 133, 135, 170
arthritis, 84, 90
assisted suicide, 56
Attleboro, 16
Australia, 85, 90
authentic, 59
autonomous, 28, 29, 45, 56–59, 70, 71, 87, 122, 130
autonomy, 5, 20, 28–30, 32, 33, 37, 38, 70, 72, 77, 119, 142, 158, 161, 163, 164
Ayd, Frank J. Jr., 59

backward-looking approach, 147
Baird, Robert M, 67, 147
Baltimore, David, 94
Bandman, Bertram and Elsie, 67
Barrington, Mary Ross, 55
Barron, D. W., 35, 36
Becker, Charlotte B., 147
Becker, Lawrence C., 147
Begin, Menahem, 9
Beitz, Charles R., 150
Belgrade Serbia, 89
Benet, Mary Kathleen, 135
Benson, Paul, 177, 178
Bernardes, Jon, 117, 158–162
Berys, Gaut, 170
Best Interest Principle (BIP), 102, 103

best interests, 23, 102, 106, 108, 110, 126, 138, 139, 156
bios, 62
biotechnology, 83, 86
Blakely vs. Blakely, 102, 103
Blank, Robert H., 95
blastocysts, 83
Blustein, Jeffrey, 105
Bock, Sissela, 57, 58
Bonnicksen, Andrea L., 95
Boston Globe, 72
brain death, 57
Brandeis, Louis, 5
Brazil, 69
Bristol County, 16
British Agencies for Adoption and Fostering, 128
British Children Act, 110
British Government, 83, 111
British Human Fertilization and Embryology Authority, 80, 84, 112
British Judge Robert Johnson, 63
British Ministry of Health, 84, 91
British Ministry of the Interior, 26
British Parliament, 27, 84
Broke, Dan, 58
Brotons, Elizabet, 106
Brown v. Board of education, 19
Buckle, Donald, 148, 149
Bulger, James, 12
Bush, George W., 174
Butler-Sloss, Elizabeth, 11, 12
Byrne, J. Peter, 48, 169

California, 99
Campbell Alistair, 71, 91, 94
Campbell, Tom, 23, 174, 179
Canada, 85
cancer, 82
capital punishment, 43, 143, 169, 170, 172, 174, 175, 179, 181
capitalist societies, 29
Caribbean Sea, 106
Carpenter, Pamela Moseley, 173
castration, 21–23, 142, 146, 156
categorical imperative, 87, 92, 130
catheterization, 61
censorship, 3, 37, 38, 44, 47, 50

Chadwick, Ruth, 78–80
Chambers, Simon, 41
Charlesworth, Max, 71, 91, 94
Cheal, David, 159, 164
child protection (CP) practices, 139
Child Welfare League of America, 136
child's right to an open future, 21, 121, 122, 125
children, 2, 15–24, 64, 87, 80, 92, 93, 97–99, 101–114, 116, 117, 119, 120, 122–140, 142, 157–166, 173, 177
chromosomes, 90
cirrhosis, 84
citizen, 1, 2, 6, 9, 11, 18, 21, 28, 30, 31, 33, 36, 66, 139, 139, 160, 161
Clements, Colleen D., 118
Clinton, William Jefferson, 8
cloning, 52, 82–91, 115
Cohen, Cynthia, 95
Cohen, Joshua, 150
Cohen, Marshall, 150
Cohen-Almagor, Raphael, 41–43
Cole, Elizabeth S., 135, 136
Colorado, 86, 91, 170
commodification, 73, 74, 99, 129, 131, 134, 136
commodities, 73, 99, 127, 134, 136, 163
common law, 169, 182
communication, 26–28, 40
community service, 142, 146, 149, 151
community, 1, 44, 48, 52, 53, 88, 94, 106, 109, 125, 126, 142, 154, 166, 167, 177
compulsory education law, 19, 20
consent to medical treatment, 23, 51
consent, 23, 24, 28, 51, 56–59, 66, 70, 73–77, 98, 99, 104, 107, 111, 116, 117, 120, 142, 146, 156
consequential, 85, 131, 141, 147
conspiracy, 10, 162
constitutions, 37, 121
Corneau, Rebecca, 16–18
Costain, Anne, 41
Cottingham, John, 147, 152
Council of Europe, 86
Court of Hadera, 146
Cragg, Wesle, 148, 150, 151
criminal law, 6
criminal records, 2

Index   185

Crist, Hetty, 111
Cruz, Oliver David, 174, 175
Cuba, 98, 106–109
Cullity, Garrett, 170
curative, 39, 142, 146, 152, 155, 156
curriculum, 19–21
custody, 15–18, 20, 22–24, 98, 99, 107, 108, 122, 131, 133
cystic fibrosis, 93

Data Protection Committee, 35
database, 2, 25, 32–34
Davis, Dena, 120
Dawkins, Richard, 85
Death Sentence, 143, 169, 170, 173–175, 179
death, 9, 12, 39, 43, 55–62, 64–67, 73, 75, 76, 79, 90, 98, 99, 106, 115–117, 119, 141, 143, 169, 170, 173–175, 179, 181
defamation, 47
dementia, 57
democracy, 6, 21, 164
Denmark, 85
Denver, Colorado, 86
deontological, 79, 86, 88
Department of Health, 111, 112
DesAutels, Peggy, 177
Desert, 154
deterrence, 141, 147
diabetes, 82, 84
diagnosis, 59, 65, 91, 94
dialysis, 61
Dickerson, Donna, 37, 70
dignity, 10, 28, 44, 56, 58, 59, 66, 73, 80, 86, 131, 132, 1134, 139, 151
diminished punishment, 172
diminished responsibility, 143, 171, 172, 176, 180, 181
discrimination, 8, 48, 88
DNA, 2, 32–34, 36, 90, 94, 97
Dolly the sheep, 83, 85, 90
domestic violence, 142, 157–162, 164–167
Donaldson Report, 83
Dowie, M., 75
Downing, A. B., 55
drug dealers, 26
Duff, R. A., 149, 171, 172

duties, 1, 2, 9, 10, 15, 19, 38, 39, 42, 50, 58, 63, 65, 78, 79, 104, 105, 107, 120, 123, 129, 139, 148, 161, 165
Dworkin, Ronald, 19, 22, 30, 58, 62, 66, 81
dysfunctional families, 166

eavesdropping, 28, 29
eavesdropping apparatus, 27
Edinburgh, 76
editors, 37, 38, 42
educational, 2
eight-cell stage, 91
Eighth Amendment, 169, 173
electronic media, 40, 43
e-mails, 2, 26
embryo dilution, 64
embryo, 17, 23, 46, 76, 91–94, 118
embryonic stem-cells (ES cells), 82
Ethics Commission of the British Department of Health, 84
Ethics Committee, 61–63, 85
European Court of Human Rights, 27
European Parliament, 86
European-Union, 86
euthanasia at the request of the patient, 56
euthanasia, 55–57, 59–61, 66
exchange value, 73, 78, 137
exploitation, 70–73, 101, 116, 129, 135
extended family, 97, 98, 101, 103–106, 117, 119, 120
Ezra, Ovadia, 87, 153

family, 57, 60, 73, 75, 92–94, 97, 98, 101–107, 109, 110, 112, 114, 115–120, 123–126, 129, 131, 133, 135, 142, 157–167
Fanconi's anemia, 91
Farber, Bernard, 162
fax messages, 28
Feinberg, Joel, 17, 21, 56, 63, 121, 122, 151
fertilization, 76, 79, 86, 87, 90, 91, 97, 111–113, 118, 119
fertilization treatment, 76, 77, 87, 91
fetuses, 2, 15–18, 64, 76, 83, 116, 129, 163
films, 38
Fingarette, Herbert, 172, 175,176, 178
fingerprints, 34
First World, 109

Fisher, John Martin, 178
Fletcher, Joseph, 67
Flew, Antony, 145
Florida, 106, 170
forward-looking approach, 147
France, 112, 123
free, 32, 59, 67, 75,
Freedman, Eric M., 46
Freedman, Monroe H., 46
freedom, 6, 15–18, 22, 27, 29, 30, 32, 38, 44, 45, 47, 48, 65, 77, 79, 84, 85, 88, 117, 130
freedom of choice 26, 153
freedom of expression, 2, 37, 38, 46
freedom of movement 26
freedom of press, 6, 8, 10, 12, 13
freedom of speech 3, 44
Frey, R. G., 57, 58
Fried, Charles, 27
Friedman Ross, Lainie, 78, 92, 119
Friedman, Leon, 46, 50

Gallagher III, Bernard J., 171
Gareth, Jones, 71, 91, 94
Garland, David, 149
Gelber, Katharine, 44, 46
Gelles, Richard J., 158–160
General Assembly, 76, 110
generosity, 74, 77, 104
genes, 81, 91, 94
genetic engineering, 52, 81–83, 93–95
genetic screening, 91
Georgia, 170
Germany, 85, 89
Giddens, Anthony, 158
Giles-Sims, 159
Gillett Grant, 71, 91, 94
globalization, 99, 127
Gonzalez, Elian, 106, 108, 109
Gonzalez, Juan Miguel, 106, 107, 109
good journalism, 42, 43
good stories, 42, 43
Governor Rick Perry, 174, 177
Graham, Gary, 174
Graham, Gordon, 5 2, 90
grandchildren, 97, 98, 102, 104, 105, 121, 124–126

grandparents, 97, 101, 106, 118–121, 123–126
Great Britain, 13, 26
grievance, 176
Group of Eight, 86
Grupp, Stanley E., 149, 153–155
guilt, 123, 145, 147, 148, 150, 153–156, 175, 177, 179

Habermas, Jurgen, 52, 81, 83, 87, 89, 91, 93, 95
Hadassah Ein Kerem, 61
Haines, Herbert H., 170
Hamner, Tommie J., 157, 158
Hampton, Jean, 150
Harding, Christopher, 151, 162
Hardwig, John, 55
Harrington, Michael, 27
Harris J, 75
Hartman, Ann, 135
hate speech, 3, 5, 44, 46, 48–50
Hawkins, Gordon, 170
Hazza, Ofra, 9
health, 2, 9, 10, 17, 28, 33, 51, 56–59, 61, 64, 65, 70, 77, 78, 90, 109–111, 118, 122–124, 126, 128
Hebrew University of Jerusalem, 44
Heginbotham, Chris, 23, 174, 179
Henting, Hans von, 165
hepatitis, 84
Hill, Donald, 16, 79
Himmler, Heinrich, 42
Hiroshima, 83
Hixson, Richard F, 7, 28, 29
Holland, Stephen, 94, 95
Homer, 165
Homicide Act, 181
homo sapiens, 53, 95
Honderich, Ted 175, 176, 179
hormonal treatments, 77, 78
Horwitz, Allan V., 172
House of Lords, 83, 84
Human Fertilization and Embryology Authority (FHEA), 80, 84, 112
human reproductive cloning, 52, 82, 84–90, 115
human rights, 5–28, 31, 37, 129, 138, 175
Hutchinson, Janice, 162, 167

Index

Hyde, Michael J., 61
hydrogen bomb, 88

IBM, 27
Illinois, 102, 103
independence, 37, 48, 106, 120, 121, 123
India, 69, 72
Indian Parliament, 72
Indiana, 170
individual, 1, 2, 4, 8, 15, 22, 24, 27–30, 32, 33, 35, 42, 70, 75, 88, 105, 117–120, 126, 131, 132, 151, 153, 177
infertility, 111, 115, 121, 125, 129
information banks, 25, 32, 33, 35
informed consent, 58, 70, 73, 77, 98, 99, 116
Inness, Julie C., 27, 29
instrumentalization, 72, 87, 89
Intelligence Departments of the British Police, 26
intelligence quotient (IQ), 93, 134, 172–174
International Covenant on Civil and Political Rights, 26
internet, 13, 26, 27, 30–32, 36, 99, 132, 134–136
intimate group, 119, 160
intrafamilial violence, 160–162, 164
intrinsic value, 38, 59, 62, 75, 78, 129, 130, 134, 137, 138
in-vitro fertilization (IVF), 83, 87, 90–93, 97, 115, 116
Ireland, Richard W., 151, 162
irreversible coma, 36, 57
Israel, 9, 10, 18, 45, 69, 122, 142, 149, 157, 158, 160, 165
Israeli Broadcasting Authority, 38, 41, 42
Israeli Supreme Court, 150
Italy, 112
Iyengar, Shant, 41

Japan, 9
Jerusalem District Court, 61
Jerusalem, 42, 152, 161
Judge Nathan Amit, 149
Judge Yaffa Hecht, 61
Justice Brennan, 173
Justice Johnson, 64
Justice Marshal, 173

Justice O'Connor, 173
justice, 12, 39, 52, 67, 76, 86, 105, 147, 148, 151, 154, 155, 167

Kamm, F. M., 75
Kansas, 170
Kant, I., 52, 79, 87, 92, 130, 131, 148
Kelly Air Force Base, 174
Kentucky, 170
Keown, John, 56, 57
Kilshaw, Alan, 132
King's College London, 83
Kingwell, Mark, 41
kinship, 101, 119, 163
Kneale, William, 180
Kolata, Gina, 86, 94
Krimmel, Herbert T., 136
Kubler-Ross, Elisabeth, 59, 60
Kushner, Thomasine, 121

LaFollette, Hugh, 55
Laird, Joan, 135
Lebacqz, Karen, 95
Lectures on Ethics, 79
Lee, Robert, 112, 125, 128
Lewis, C. S., 149, 154, 155
Lewis, David, 148
Lex Talionis, 17, 148, 150
life threat, 62
life, 5, 7, 10, 13, 14, 17, 20–22, 28, 31, 32, 36, 47, 48, 51, 52, 55–67, 69, 70, 73–76, 78, 80, 85, 87–92, 94, 95, 97, 99, 103, 109, 113–117, 120–122, 124, 125, 127, 133, 138, 143, 159, 161–163, 166, 175, 181
life-saving procedures, 62
Locke, John, 46
London, 83, 111, 162
Lord Woolf, 12

Ma'ariv, 42
Mannes, Maria, 59
Mansfield, Harvey, 31
marital violence, 158
market inalienable, 130, 131, 134
market rhetoric, 128, 130–132, 134, 139
Maryland, 170
Massachusetts, 16

Mcge, Glen, 93
McNaghten Rules, 179
McQuail, Denis, 11
McVeigh, Timothy, 39, 43
media, 2, 3, 7, 9–11, 24, 37–41, 44, 48, 132
medical confidentiality, 2
Megivern, James J., 175
Mellors, Colin, 11, 14, 27, 34
mens rea, 143, 178–181
mental retardation, 21, 23, 169, 173, 175
mental state, 61, 62, 143, 175, 181
mentally ill, 143, 169–172, 174–176, 179
mentally retarded, 2, 15, 22–24, 70, 122, 141, 143, 169–182
mercy death, 51, 56–61
mercy killing, 51, 56, 57, 60, 61, 63, 65, 66
mercy, 41
MI5, 26
MI6, 26
Miami, 106
Middle East, 160
Mill, John Stuart, 45–49
minors, 2, 15, 70, 122, 123, 134
Missouri, 133
Missouri Supreme Court, 133
MIT, 94
mitigated punishment, 142, 146
Mitigating Circumstances, 170, 181
mitochondria, 84
Model Penal Code, 170, 179
monitoring devices, 27
Moore, D., 74
Morgan, David, 164
Morgan, Derek, 112, 125, 128, 129
Morris, Cynthia, 169, 173, 182
Morris, Herbert, 154, 175, 176
Morris, Norval, 148, 149
motor neurone disease (MND), 83
multiple sclerosis, 84
multipotent stem-cells, 82
Munro, Colin, 42
Murphy, Jeffrie G., 147, 148
Murphy, Timothy, 70, 94
Murray, Thomas H., 111, 116, 117, 130, 131, 136
mutuality, 101, 131

Nagel, Thomas, 6
Nash, Jack and Lisa, 91–93
Nasif, Kenneth P., 16
Natanya court, 146
national emergencies, 25, 26, 31, 32
Nazi, 42
Nebraska, 170
Negative Eugenics, 53, 91, 94
neuro-degenerative diseases, 84
New Mexico, 170
New York, 170
New Zealand, 111
Nickel, James, 10, 25, 26, 31
non-voluntary active euthanasia (NVAU), 56
nuclear family, 97, 98, 101
nuclear replacement (CNR), 83
Nussbaum, Martha C., 72, 85, 86

O'Donovan, Katherine, 112
O'Neill, Onora, 77, 115, 118, 123–125, 135
O'Shea, Kathleen, 170
objectification, 72, 79, 129
obscenity, 47
offspring, 22–24, 86, 87, 97, 107, 110–114, 116, 119, 125
Oklahoma city, 39
omnicide, 88
On Liberty, 45
one's right to be let alone, 2, 3, 5, 25, 33
oppression, 6
Orwell, George, 30
Oslo Accord, 42
osteoporosis, 84
Ostheimer, John M., 59, 60
Ostheimer, Nancy C., 59, 60
Ought Implies Can, 21, 85
ovarian cancer, 78
ovum, 52, 76–78, 80, 93

Pacific Reproductive Services, 112
Page, Edgar, 16
paid adoption, 99, 132, 134, 135, 138
paranoid schizophrenia, 174
Parens Patriae, 15, 20, 21, 122, 129, 138, 139, 165
parental autonomy, 161
parental rights, 2, 16, 18, 20, 86, 98, 105, 108–110

parent–child violence, 159
parenthood, 97, 98, 102, 109, 110, 112–114, 116–120, 124, 128
parents, 2, 16–18, 20–24, 64, 65, 87, 90–93, 97–99, 101–129, 133–140, 159–164
Parkinson's, 82, 84
passive euthanasia, 60, 61
patriotism, 11
Patterson, Kelsey, 174
pedophiles, 26
Pence, Greg, 5
Penry, John Paul, 172
persistent vegetative state, 57
personhood, 30, 32, 93, 120, 131, 134, 138, 139
Petrinovich, Lewis, 72, 75
Philippines, 69
pictures, 38, 39, 42
plea bargains, 142, 145–147, 152, 155
pluripotent ES cells, 82, 83
police, 6–8, 16, 26, 28, 33, 94, 140, 158, 162, 167
Prado, C. G., 56, 67
preemptive suicide, 56
pregnancy, 17, 64, 117, 129
preimplantation genetic diagnosis (PGD), 91, 93
President Bill Clinton, 85
Presidential Ethics Commission, 85
press, 5, 6, 8–13, 40
privacy, 1, 2, 5–11, 13, 14, 26–36, 39, 105, 112, 113, 161, 163, 166
privilege, 3, 8, 12, 13, 38, 103, 161
procreation technologies, 114–116
procreation, 22, 98, 115, 118, 119, 121, 125
procreative rights, 22, 135
prognosis, 59, 65
proportionality principle, 141, 142, 146, 148
prosecution, 7, 142, 146, 150, 152, 155, 156, 166, 167
prosecutor, 6
Prosser, William, 7, 8
prostitution, 99, 127
proviso, 25
public figures, 2, 6–8
public interest, 2, 6, 8, 9, 11, 12, 38, 39, 65, 82

public's right to know, 6–11, 13, 39
punishment, 12, 13, 17, 43, 77, 141–143, 145–162, 164, 165, 167, 169–182

Rabin, Yitzhak, 42, 43
Rachels, James, 60, 61, 63, 153
racism, 42
Radin, Margaret Jane, 128, 130, 131, 134, 138, 139
radio channels, 40
Ramsey, Paul, 74, 118
rapist, 38, 43, 146, 149, 150, 164
ratings, 40, 42, 44
Rau, Johannes, 89
Ravizza, Mark, 178
Rawls, John, 147
reform, 152
rehabilitation, 12, 142, 145, 146, 153–156
Reich, Warren T., 118
Reiman, Jeffrey H., 30, 31
Reno, Janet, 109
repayment, 147
repetitive DNA, 90
Report on the Committee on Mentally Abnormal Offenders (the Butler Report), 171
reporters, 37
reproductive autonomy, 115, 125
responsibility, 11, 15, 24, 36, 40, 44, 59, 90, 94, 95, 107, 119, 120, 122, 123, 126, 131, 138, 139, 143, 154, 165, 170–172, 175–178, 180, 181
restoration, 133
retribution, 147, 148, 150
Rhodes, Rosamond, 121
right to bodily integrity, 23, 71, 72, 92
right to choose, 115
right to privacy, 5–11, 13, 14, 26, 27, 29–34, 36, 39
right-based theories, 86
Roberson v. Rochester Folding Box Co., 29
Robertson, John A., 125, 136
Robinowitz, Carolyn B, 162
Robison, Larry, 174
Rosenbaum, Stuart E., 67, 147
Rosenthal, Miriam B., 111
Roslin Institute in Edinburgh, 83, 89, 90

Ross, Jacob Joshua, 103, 105–107, 112, 123, 163
Rostankowski, Cynthia, 67

Saatkamp, Herman, 93
Sadurski, Wojciech, 48, 49
Safeguard Guaranteeing Protection of the Rights of Those Facing the Death Penalty, 169
same-sex couples, 102, 118
San Antonio, 174
San Diego, 99, 132
San Francisco, 112
Satterwhite, John, 174, 175
Schabas, William, 169
Schauer, Frederic, 47
Schloendorff v. Society of NY Hospitals, 70
Schoeman, Ferdinand D., 7, 8, 11, 28, 29
scientific community, 52, 53, 90, 91, 93, 94
Scotland, 76, 89
secrecy, 6–11, 14, 27, 34, 35, 89, 162
September 11, 2
sex crimes, 142, 145, 146, 149–152
sex offenders, 141, 142, 145, 146, 149, 151
sexual orientation, 48, 93
Sharon, Ariel, 7
Shaw, Christopher, 83
Shue, Henry, 21
Siamese twins, 51, 63
sickle-cell, 84
Simmons, John A., 150
Singapore, 90
single-parent family, 117
Sinsheimer, Robert L., 95
Sky News, 134
slavery, 99, 127, 138
slippery slope, 58
Smoker, Barbara, 55
Smolla, Rodney, 48
social solidarity, 74
Spain, 85
sperm, 77, 78, 80, 83, 93, 97–99, 111–117, 119, 121, 124, 125
Spurlock, Jeanne, 162
St. Mary's Hospital in Manchester, 64
Steinmetz, Suzanne K., 158–160
stem-cell research, 52, 52, 82–84, 93, 95
stem-cells, 53, 82–84, 91

sterilization, 23
sterilize, 22
Straus, Murray A, 158–160
Sunstein, Cass R., 85, 86
surplus value, 137
surrogacy, 121
surrogate, 119, 121, 123, 125
Sussman, Marvin B., 158
Sweatt v. Painter, 19
Sweatt, Herman Marion, 19
Sweden, 111, 117
Switzerland, 85, 112
sympathy, 74

Teichler-Zallen, Doris, 118
Tel Aviv District Court, 146
telomeres, 90
Ten, C. I., 150, 171, 172, 178, 179
Tennessee, 170
terror, 39, 42
terrorism, 2, 25, 32, 34, 160
terrorists, 26, 31
Texas courts, 143, 173, 175, 178, 181, 182
Texas, 170, 173, 174, 177
The American Law Institute, 170, 179
The European Convention on human rights, 26
The Hague Convention on the Civil Aspects of International Child Abduction, 107
The Hague, 107
The Humanitarian Theory of Punishment, 154
The Liberty of Thought and Discussion, 45
The Precautionary Principle, 90
The Sacred, 62
therapeutic cloning, 53, 83, 84
Third World, 72, 107, 109, 128, 135
Thiroux, Jacques, 56
Thomasma, David C., 121
Thompson, Robert, 11
Thomson, Judith Jarvis, 29
Tinder, Glen, 46
Titmuss, Richard, 80
Tizard, Barbara, 127, 133
Todd, R. W., 27
toleration, 48
totipotent stem-cells, 82
Trager, Robert, 30, 37

Index

Tranda, W., 132
transplantation, 51, 69–71, 73–77, 83, 84, 89, 92
Trost, Jan, 118, 158
Troxel vs. Granville, 102, 103
Turner, Pauline H., 157, 158
TV, 2, 39–41
twilight zone, 132, 166
tyranny, 6

UK Office of Population Censuses and Survey, 128
UN Commission on Human Rights, 72, 174
UN Committee on Crime Prevention and Control, 169
UN Convention on the Rights of the Child, 110, 139
UNESCO, 86
Universal Declaration of human rights, 26
University of Newcastle, 84
University of Texas, 19
university professors, 37
unmarried mothers, 117
Urban Walker, Margaret, 177
US constitution, 173
US Supreme court, 21, 102, 103, 109, 169, 173
US, 9, 10, 34, 41, 98, 106, 158, 173
use value, 73, 137
utilitarian, 69, 75, 86, 91, 141, 147, 160

values, 6, 8, 18, 29, 32, 35, 37–39, 48, 52, 55, 57–59, 62, 63, 66, 67, 69, 73–75, 78, 80, 81, 85, 86, 117, 120, 125, 129–131, 134, 137–139, 142, 146, 147, 150, 151, 154, 161, 164
Van Creveld, Martin, 44–49
van den Daele, W., 89

Velasquez, Manuel, 67
Venables, Jon, 11
Vila, Bryan, 169, 173, 182
vilification, 44
visitation rights, 102–104, 106, 126
voluntary active euthanasia (VAU), 56
von Hirsch, Andrew, 149
Von Humboldt, Wilhelm, 47

Walker, Nigel, 150, 151
Wallace, R. J., 170, 171, 179
Walsh, Paul F., 16
Warnock, Mary, 123
Warren, Samuel, 5
Washington, 170
Wasserstrom, Richard A., 17, 27, 30, 177
Wecker, T., 133
Wellman, Carl, 32, 63
Werner, Richard, 17
Westin, Alan, 11
white-collar criminals, 26
Whiting, Raymond, 55
Wilkinson, Stephen, 69, 71–73, 76, 80, 129–131
Wilmut, Ian, 83
Winfield, Richard Dien, 106, 164, 165
Woosley, A. D., 63, 129
Wootton, Barbara, 177, 181
World Health Association (WHA), 76

Yedioth Ahronoth, 42
Young Committee, 35
Young, John B., 11, 27, 35

Zimring, Franklin E., 170
zoe, 62
Zoloth, Laurie, 95
zygote, 82